The Public Role in the Dairy Economy

Also of Interest

Dairy Science Handbook, Volume 15 (International Stockmen's School Handbooks), edited by Frank H. Baker

Modeling Farm Decisions for Policy Analysis, edited by Kenneth H. Baum and Lyle P. Schertz

*Science, Agriculture, and the Politics of Research, Lawrence Busch and William B. Lacy

Livestock Behavior: A Practical Guide, Ron Kilgour and Clive Dalton

Animal Health: Health, Disease and Welfare of Livestock, David Sainsbury

Animal Agriculture: Research to Meet Human Needs in the 21st Century, edited by Wilson G. Pond, Robert A. Merkel, Lon D. McGilliard, and V. James Rhodes

Animals, Feed, Food, and People: An Analysis of the Role of Animals in Food Production, edited by R. L. Baldwin

Zeo-Agriculture: The Use of Natural Zeolites in Agriculture and Aquaculture, edited by Wilson G. Pond and Frederick A. Mumpton

*Available in hardcover and paperback.

Westview Special Studies in Agriculture Science and Policy

The Public Role in the Dairy Economy: Why and How Governments Intervene in the Milk Business
Alden C. Manchester

All over the world, governments play a part in the milk business for compelling economic reasons and not, as many assert, just because dairy farmers are numerous and organized. This book examines the role of federal, state, and local governments in the dairy economy of the United States, where major public involvement in industry began during the Great Depression. Dr. Manchester considers the conditions in the 1930s that led to government involvement, the changes that have occurred in the industry and the public role since then, and the prospects for the 1980s and beyond. He also analyzes possible alternative public dairy policies for the present and the rest of the decade. Many things have changed, points out Dr. Manchester, but the fundamental conditions that led to public involvement in the dairy industry still exist.

Dr. Manchester is senior economist, National Economics Division, Economic Research Service of the U.S. Department of Agriculture. He was in charge of dairy marketing research in the USDA from 1961 through 1973. His other publications include *Organization and Competition in the Midwest Dairy Industries* (1970), *Dairy Price Policy: Setting, Problems, Alternatives* (1978), and *Market Structure, Institutions, and Performance in the Fluid Milk Industry* (1974).

The Public Role in the Dairy Economy

Why and How Governments Intervene in the Milk Business

Alden C. Manchester

Routledge
Taylor & Francis Group
LONDON AND NEW YORK

First published 1983 by Westview Press

Published 2019 by Routledge
52 Vanderbilt Avenue, New York, NY 10017
2 Park Square, Milton Park, Abingdon, Oxon OX14 4RN

Routledge is an imprint of the Taylor & Francis Group, an informa business

Copyright © 1983 Taylor & Francis

Library of Congress Cataloging in Publication Data
Manchester, Alden Coe, 1922-
      The public role in the dairy economy.
      (Westview special studies in agriculture science and policy)
      Bibliography: p.
      Includes index.
      1. Milk trade--Government policy--United States. 2. Milk trade--
United States--Price policy. 3. Dairy products--United States--Price
policy. 4. Agricultural price supports--United States. I. Title. II. Series.
HD9282.U4M324   1983   338.1'873   83-6957

ISBN 13: 978-0-367-29540-0 (hbk)
ISBN 13: 978-0-367-31086-8 (pbk)

Contents

Part 1. PROLOGUE

Part 2. INDUSTRY STRUCTURE

Tables and Figures

FIGURES

Foreword

It is widely believed that the author owes it to his readers to communicate to them "where he is coming from." Although I lack confidence in the possibility of communicating such a matter, for those who subscribe to this belief I offer the following. I lack confidence in Right Answers. I do not believe that there are any eternally correct solutions for human problems in economic fields or elsewhere. Especially, I lack confidence in simple answers. As Winston Churchill is alleged to have said, "For every complex human problem, there is a solution which is simple, straightforward, and wrong."

The hope for humanity in matters economic is first to understand how and why the system works. Perhaps then we can begin to understand what constitutes the more prudent course in economic policymaking. Prudent, in this case, means the course that will do the least damage to the functioning of the economic system.

I lack o'erweening faith in what are euphemistically called "free markets," which work magically in the textbooks but quite differently in the world "out there." On the other hand, I find little reason to conclude that governments are all-wise and efficient in all they do. What we find in practice, of course, is universally some mixture of private markets and governmental participation. In a modern industrial economy, it seems unlikely that we could operate without both.

In fact, governmental participation is required to create markets. It is the existence of rules defining property rights, legal tender, and other matters that makes a market possible.

The perfectly competitive markets of textbook theory never existed and could not be created. They are a figment of the theorist's imagination, convenient for exposition and helpful in understanding the economic forces involved. They are not useful guides to public policy. There is no such thing as an unregulated market, much less an unregulated market that is perfectly competitive in the textbook sense. An intricate set of regulations would be needed to move even a little bit in the direction of a perfectly competitive market. Thus, regulation is not just a means of dealing with market failure—without regulation in the broadest sense there would be no market. (See Shaffer, 1979, p. 722.)

In such a situation, the appropriate question for public policy is: what is the blend of public and private activities that will stimulate better performance of a mixed system? The concept of "market failure," which holds that governments should intervene only when the private economy fails, is based on the right idea. But, by stating the argument in such apocalyptic terms, its premise reaches the conclusion desired by its promoters: governments should only intervene when there is a complete breakdown in private markets. Complete breakdown, while not impossible, is not the question. The question is: would a measure of government participation improve performance?

I am also in profound disagreement with the point of view that "all you need to know is microeconomic theory." I strongly believe that any analysis that does not take into account economic, biological, and other characteristics of an industry will miss the mark by a wide margin. Fortunately, in the case of the dairy industry there is a sixty-year history of economic analysis that provides a solid base on which to build.

Alden C. Manchester

Acknowledgments

My greatest debt is to the many researchers on the economics of the dairy industry who have labored in these vineyards for many years. Because of their labors, we can speak with confidence as to how things are and how they got that way. Many of their works are listed in the References. I apologize to any who may have been inadvertently overlooked.

My second greatest debt is to my reviewers:

Richard A. King of North Carolina State University
Emerson M. Babb of Purdue University
Milton A. Hallberg of Pennsylvania State University
Joel L. Blum of the Agricultural Marketing Service
Robert W. March of the Agricultural Marketing Service
  (retired)
Richard F. Fallert of the Economic Research Service

They have made me do a better job, both in covering topics that I might otherwise have ignored and in making clearer what I was trying to say. I am more grateful than I can say for their help.

My thanks also go to those who have labored over the years to turn my spoken and scribbled words into legible copy. They include Patty Beavers, Lorraine Cate, Marionetta Holmes, and Betty Acton.

A.C.M.

Part 1

Prologue

1
Introduction

Government is more involved in the production and marketing of milk and dairy products than in most agricultural products. Federal, state, and local governments have all played roles, with the emphasis shifting toward the national government over the years, as it has in most economic activity. This is not a uniquely American trait--the government of every major milk-producing nation in the world intervenes in one way or another. This book considers the characteristics of the dairy economy that lead to governmental involvement, the form of such intervention, and the performance of the resulting mixed public and private system.

The intervention of governments in the economy arises from a wide variety of goals, ranging from income redistribution--a primary goal of price and income policy--to specific social goals arising from externalities--such as the maintenance of environmental quality --to those not market-oriented at all--such as prevention of disease. Economic activity is affected by government in many ways. Resources are allocated by the market, but always under a set of rules both private and public. Public involvement in markets typically occurs when private rules no longer are effective in accomplishing perceived public goals--when the performance of those markets violates the values of buyers, sellers, or third parties. Then governments intervene with the intent of improving market performance--i.e., moving performance closer to the perceived goals.

Major involvement of Federal and State governments in the dairy economy began during the Great Depression, so this book begins then after a brief look at what had gone before.

The economic and biological characteristics of the dairy industry will be examined, the conditions leading to Federal intervention in the thirties described, the changes in the dairy industry and in public programs since then analyzed, and the conditions which would evolve in the eighties without Federal regulation considered.

Broadly speaking, the major reason for the introduction of Federal dairy programs in the thirties was the economic collapse caused by the Depression. This involved both price levels--too low to prevent bankruptcy--and stability.

The dimensions of stability will be explored. It is typically understood by economists to refer to variations in prices between seasons and between years. But the instability which led to classified pricing by cooperatives in the twenties and by the Federal government in the thirties was related more to the instability which arose from non-uniform prices among handlers, i.e., to instability in competitive relations resulting in some handlers paying lower prices to producers than those paid by others. This type of instability later came to be associated with the term <u>orderly marketing</u>.

In production, the technology and organization of the production unit have changed drastically. However, the basic biology has not changed. Production is still continuous and subject to a biological lag.

Changes in processing and marketing have also been significant, but many of the basic characteristics remain. Changes in public programs have been numerous and far-reaching. Have they kept pace with changes in the industry?

A major element is change in cooperative power. Its dimensions will be considered.

A major question is: Given the changes in the dairy industry since the thirties, what would be the effects of eliminating Federal dairy programs? This will be examined in terms of the three major Federal programs—price supports, market orders, and policy regarding cooperatives.

The principal effect of the price support program is to provide a floor under milk prices at the low point (in terms of prices) of the cow cycle. The analysis will consider the effects of removing the floor—would prolonged periods of low prices result? Or would a major cow cycle return?

Examination of the effects of discontinuing Federal marketing orders and comparable State regulation involves two stages. The first will try to isolate the economics from the institutional phases of the question. It will focus on the economics behind the classified pricing system.

The second stage looks at the institutional setting, focusing on stability or instability in competitive relations. This is, of course, primarily a question of cooperative conduct and power. It will go back to the bases of cooperative power to see how they might be affected by the demise of market orders. A variety of modifications in policy instruments, other than abolition of the programs, are also considered.

Policymaking is preeminently a political process with all of the tradeoffs, consensus-building, and coalition-construction which that implies. The give-and-take of both Executive and Legislative policymaking involve other agricultural commodity programs, international trade policy, environmental policy, antitrust policy, tax policy, energy policy, and national and international monetary policy. This book is not about the political process but about the economics of the dairy industry and of public involvement in it.

2
Characteristics
of the Dairy Industry

The dairy industry in the United States is not like other agricultural and nonagricultural industries. In many ways, it is unique. That does not make the dairy industry more or less worthy than the others. It does mean that the differences must be understood because they play a significant role in determining the effects of different public policies towards the dairy industry. Policies which will achieve a given set of objectives for a crop (say potatoes) will have entirely different results in the dairy industry. This chapter discusses the major characteristics of the dairy industry and the ways in which they differ from those of other agricultural industries.

PRODUCTION

Specialized inputs--The production of milk requires specialized inputs--milk cows, buildings, machinery, and skilled management. All of these have little or no value in other uses, i.e., once in place, the assets are fixed. In contrast, land and some of the machinery used for crops (tractors, tillage equipment, fertilizer spreaders, etc.) can be used for a variety of products. This accounts for the shifts in acreage to different crops on a given farm in response to anticipated returns.

Rate of output--Choice of rate of output is severely limited for milk production. Relatively small adjustments in output can be made in the short run by changing feeding rates, but major adjustments require changes in cow numbers. Downward adjustments can be made rapidly by culling dairy cows. Increases in output in response to higher prices are limited in the short run by the number of replacement heifers available. Large adjustments require raising a higher proportion of female calves (slightly less than one-half per milk cow) for two years. Thus, major increases in production are subject to a multiyear biological lag.

The closest resemblance to milk production is found with eggs, but the biological lag is much shorter--production can be increased in seven months from the time additional eggs are placed in the hatchery until the first eggs are laid by the new pullets.

<u>Entry or exit</u>—Entering or leaving milk production involves major shifts in capital structure. Investments of substantial capital are required for entry. Shifts among crops on a given farm require relatively little capital.

<u>Inspection</u>—Production and holding facilities on the farm are subject to frequent inspection by health authorities. The necessity for such inspection to protect the public health does not exist for any other farm product.

HARVESTING

Harvesting of any one crop is a single-time thing. A farmer's harvest may be spread over several weeks or months by a series of plantings, as is typically done for fresh market vegetables, or by using varieties with different maturity dates—both fruits and vegetables—but one plant produces once a year, although some fruits can be stored "on tree" for an extended period (e.g., oranges).

In contrast, milk is "harvested" twice a day every day of the year. The daily "harvest" of eggs is the only close counterpart.

Milk that is "harvested" must move off the farm. Up through the forties, milk for fluid use had to move every day, since refrigeration technology (on the farm and in transport) could not maintain an acceptable level of quality longer than that. Today, storage on the farm for more than 48 hours, while technically possible, is extremely costly and discouraged by health authorities. No other product must move as rapidly or as regularly. Eggs can be held for a week, grains for several years. A few fresh vegetables and fruits which decline in quality fairly rapidly must be moved within 24 hours to maintain quality, although not because of the possibility of contamination.

The market-clearing value of a unit of output varies among producing units (by location and product characteristics) and from day to day on any given farm. The value on any given day is a function of the supply available that day and the demand, both of which vary randomly and with some systematic characteristics.

MARKETING

Farm-to-processor assembly of raw milk is costly and must be scheduled well in advance of "harvest" to prevent quality deterioration or loss of saleable product.

Milk is used in many products. It is routinely disassembled into ingredients and then reassembled into a wide variety of different products; many become ingredients in the manufacture of other food products. There is some similarity in this characteristic with eggs and fats and oils.

Conversion of milk from raw product to the final product forms entails use of capital-intensive, high-technology facilities that are subject to significant economies of scale.

Storage to balance milk output with sales is feasible only after processing, except in the briefest time span.

The services involved in marketing milk, especially fluid milk, are numerous and costly. The brief storage period for milk in fluid form creates much of the need for services.

Demand for fluid milk products is more continuous than for most foods, with milk being consumed every day in the majority of milk-using households. Purchase patterns have become lumpier than consumption with the shift from home delivery to store buying.

UNIQUE BUYER-SELLER RELATIONSHIPS
IN THE DAIRY ECONOMY

The dairy farmer has a sole customer who
... must take the total output of the herd
... loaded out at a prearranged time of day
... every other day of the year without fail
... at a mutually agreeable price or prices.

The processor has many customers who
... make a variety of products that
... sell at different but reasonably stable
    prices with
... each item fabricated daily from
... the uncertain volume of raw milk
    purchased from
... a fixed set of dairy farmers
... who are paid monthly at an average price
... based on the mix of final products sold.

INHERENT INSTABILITY

Instability is inherent in fluid milk markets. Production varies seasonally largely for biological reasons and from day to day—neither type of variation is coordinated with changes in demand. The following models will help to make this clear. The first considers only seasonality of production. 1/

Take a simple case of one central city in the middle of a milk-producing region. Producers ship to the city if the city price less shipping cost exceeds the manufacturing price less transportation (the manufacturing plant being nearby). Plants in the city adjust their prices seasonally so that, at any time, just enough producers ship to the city to meet the city's fluid needs. This defines three separate rings. All of the annual milk output from the inner ring is used for fluid products. The second ring is defined as those points between the seasonal minimum and maxiumum radii needed to serve the fluid market. Milk from this ring that is not needed for the fluid market is manufactured. All of the milk from the third ring is used for manufacturing.

1/ This section draws heavily on Miller (1982a).

This solution would appear optimal. Milk at all points flows, in response to price signals, to the highest-valued use at any given time. However, this solution is not stable because of the unequal distribution of costs and revenues among producers or plants.

Prices received in the inner and outer rings are straightforward. In the inner ring, producers receive the average city price minus transportation cost for all their milk. Similarly, the price in the outer ring is determined by the value of manufactured products minus the cost of manufacturing and transportation. Manufacturing plants in the middle ring will utilize all of the ring's milk during the heaviest surplus season and none of it during the tightest conditions. Because of this variability, their fixed costs per unit will be sharply higher than those of plants in the third ring. The price paid by these plants for milk for manufacturing must be below the price in the third ring if these plants are to be viable. The price received by a second-ring producer is the weighted average of the prices received by shipping to the city and the second-ring manufacturing price. The average price near the inner boundary of the second ring is just slightly below the average city price minus transportation. Near the outer boundary, the price is below the third-ring price.

This provides incentives for producers near the extreme of the second ring to enter an agreement with manufacturing plants to ship all their milk to them and none to the city. The price rise resulting from the consequent city milk shortage provides a windfall to producers further in and extends the outer boundary of the second ring. Toward the inner portion of the second ring, producers may organize to ship all of their milk to the city, whether or not it is needed. Such a system creates inherent price and organizational instability and increases market transportation costs.

Beyond the question of unstable prices and fluid milk supplies, a serious equity problem arises. Variability arises from conditions in the market as a whole, yet the costs of handling it are borne only by second-ring producers (particularly those at the fringe). A stable solution can be found only by a transfer of revenue from first-ring to second-ring producers. There is one major danger to such a compensation scheme. If producers at the fringe become completely indifferent to where their milk goes, all incentive to provide milk to the city as needed disappears. Compensation must be contingent upon actual performance to be completely effective.

No market ever conformed precisely to this model. However, a closely analogous situation was common. Some plants or cooperatives performed the balancing function, with others receiving the benefits without incurring much cost. The favored way of avoiding balancing cost was to attract producers in manufacturing areas only as needed. Part of such a plant's balancing costs were shifted to balancing plants within the market and part was shifted to manufacturing plants outside the market.

The preceding model deals with variation (primarily seasonal and weekly) which can be handled by varying the flow of milk from the fringe areas to the city. An analogous situation develops cyclically as the overall milk supply moves from ample to tight. Producers and their organizations who supply a market at one time can see no reason why they should be cut off when production increases. Daily variability must be handled by having a pool of reserve milk close enough to the city to reach city plants on any given day. Assume that a manufacturing plant located near the inner extreme of the second ring is within the same-day shipping radius. This plant will receive a supply of milk which is available to city plants on call. The milk not needed by the city plants is manufactured. The size of this supply is determined by the difference between the minimum and maximum quantities which might be needed by city plants on that day.

Costs of handling daily variability arise from direct costs of coordinating shipments, added transportation, and extreme variability in manufacturing the reserve. In this case, direct costs of coordination are minimized by concentrating the daily balancing function in a single plant. However, there remain positive costs in arranging shipments so that each individual city plant receives just as much milk as it requires on that day. Such costs include rerouting shipments, maintaining standby transportation capacity, reloading, maintaining information flows, and others.

Having enough milk at the manufacturing plant to meet the maximum possible fluid demand requires that more milk than is needed (on average) be shipped part of the way to the city. Suppose that for a given day, the quantity of milk expected to be used by city plants requires half of the second-ring output. However, three-fourths of the ring's production will be needed to cover the maximum possible demand. In this case, the expected cost is the additional transportation cost associated with shipping a fourth of the ring's milk to the inner manufacturing plant. If no manufacturing plant lies within the same-day shipping radius, there would be the additional cost of shipping the unneeded milk back to a manufacturing plant.

During the heaviest surplus periods, most of the daily reserve can come from that plant's normal deliveries. During the rest of the year, other manufacturing plants must release a part of their deliveries to provide the reserve. This heightens their seasonality and, therefore, their costs of seasonal balancing. However, the cost burden falls most heavily on the plant maintaining the daily reserves. It must have manufacturing capacity for all the milk produced by its producers during the heaviest surplus period plus any additional daily reserve and it will have many days with no milk to manufacture. Fixed cost of products manufactured will be very high. In addition, costs that are variable when dealing with seasonal variation, such as some labor costs, become fixed at high levels per unit because of the inability to adjust input quantities to varying daily levels of output.

Seasonal variability has tended to diminish over time. However, weekly and, particularly, daily variability have increased as home delivery sales were replaced by sales through stores. The normal intraweek pattern is pronounced in most markets. The variability around this normal daily pattern is severe. Costs of daily balancing are inherently very difficult to quantify and exhibit substantial economies of central coordination. Under these conditions, only a monopoly coordinator can achieve a solution that is equitable to all producers and approaches the least-cost solution.

IMPLICATIONS

Some of the results of the above characteristics of milk production and marketing are--

o The number of processors is very small compared to the number of producing units.

o Long-term commitments by producers and processors are economically advantageous to both parties.

o Terms of exchange must take into account both security and equity, while permitting appropriate responses to changes in market conditions.

o Producer groups often are able to reach more favorable terms than individual producers.

o A public agency may serve as an arbitrator between buyers and sellers, providing a measure of stability which they cannot provide for themselves.

o Instability is inherent in fluid milk markets. Institutions must be created to deal with it.

3
The Dairy Industry and Public Policy Before 1933

For most of the life of this country, most of the people raised a very high proportion of the food they ate. Only in the past century, and particularly since World War I, have most of the people come to rely on food produced by others. In the Colonial period, nearly everyone lived in rural areas. In 1790 when the first census was taken, nearly 95 percent of the population lived in rural areas, places with populations of less than 2,500 persons. Less than 3 percent of the population lived in cities of 10,000 or more. All through the nineteenth century and up to World War I, a majority of the population lived in rural areas. A high proportion of these rural residents kept one or more cows and supplied their own needs for dairy products. Most of the others could obtain milk, butter, and cheese from their neighbors whose cows produced more than the family needed.

But the cities, few as they were, also had to be fed. Boston-- the biggest city until well into the nineteenth century--was particularly poorly situated for the keeping of cows, even before they were squeezed out by people. Pasture was scarce and not very good and Boston very soon became dependent on nearby towns for dairy products. Settled in 1630, by 1687 there were only 177 cows and nearly 7,000 people. In 1742, there were 141 cows and 16,382 persons. Milk crossed the river from Charlestown and Cambridge and undoubtedly came overland from other towns. Many of the peddlers who traveled the streets of Boston selling milk were certainly farmers from nearby towns or members of their families. It seems likely that others were specialized peddlers who bought milk from farmers and sold it in town, although we have no evidence. (See Friedmann, 1973.)

Farm-made cheese was an early item of commerce. By the 1740s, cheese was exported in some quantity, especially to the West Indies, and shipped by water to New York and other cities--all cities were at the water's edge in those days.

Milk production was a minor sideline for most farmers, although some farmers who supplied milk for city markets must have been somewhat more specialized dairy producers, even in Colonial times. But in South County, RI, farms were large and specialized. In the 1750s, William Douglass described the largest farm as milking 110 cows, cutting 200 loads of hay, and making 13,000 pounds of cheese and some butter in a year (Douglass, 1755). These atypical large farms were worked with slave labor, both black and Indian (Russell, 1976).

As the frontier moved westward, first New England and then New York and Pennsylvania ceased to be general farming areas and specialized in dairy farming. The construction of the railroads in the second quarter of the nineteenth century drastically altered the potential markets of dairy producers. With quicker transport and growing urban markets, specialized production for use in fluid milk began to develop fairly close to cities and for butter and cheese in more distant areas. Boston was the first to use railways. By 1842, when the first shipment came into New York City on the Erie, Boston was receiving from many points within 10-30 miles (Gates, 1960, p. 238). Not until refrigerated cars came into use in the 1870s did shipments of more than 100 miles become feasible (Merritt, 1965, p. 12).

Increased specialization in milk production brought a 40 percent increase in annual milk output per cow between 1850 and 1910. Nearly 60 percent of the increase was due to a longer milking season (300 days in 1910 compared to 237 in 1850) and 40 percent to better feeding and better techniques (Bateman, 1968).

Factory production of butter and cheese began in the 1840s, grew some in the fifties, and then accelerated, first in response to Army purchases during the Civil War. Farm cheese production was quickly replaced by factory output--the 1849 farm production was the peak. Farm butter production, which required less skill than cheese, continued to increase through the nineteenth century and peaked in 1908.

Perhaps 6 percent of all farm milk was sold in 1849--plus a large but unknown portion of the farm-made cheese (table 1). The share sold increased to a fourth in 1879, a half in 1909, and almost three-quarters in 1929. Use in manufactured products declined from three-fourths in 1849--nearly all farm cheese and butter--to 60 percent in 1899 and 57 percent in 1929. Sales for fluid use rose from 6 percent of production in 1849 to 28 percent in 1929.

Growth in the fluid milk business brought many changes. Farms soon came to specialize in production for that market and milk dealers went into business, first in the large cities and later in middle-sized centers. By 1880, most cities had a core of large milk dealers and a large group of small ones. In Boston in 1897, seven "contractors" (dealers) handled about three-quarters of the milk and several hundred farmers and peddlers the remainder (Whitaker, 1898).

Thus, the atomistic fluid milk marketing system of the earlier years where farmers sold milk directly to consumers or sold to peddlers who dealt with consumers was replaced in the last quarter of the century by a much different structure with hundreds or thousands of dairy farmers selling to a handful of large dealers and a dual system that still retained the atomistic characteristics of an earlier day.

Table 1 – Milk production and use, 1849–1929

Year	Production			Used on farms				Sold by farmers		
	Nonfarm	Farm	Total	Fed to calves	As milk and cream	For butter and cheese	Total	To consumers	To milk dealers	To manufacturers
						— billion pounds —				
1849	0.6	10.2	9.6	0.3	1.7	7.6	0.6	0.4	0.2	*
1859	.8	14.4	13.4	.4	2.3	10.7	1.1	.5	.5	0.1
1869	1.0	17.6	14.6	.5	2.8	11.3	3.0	.8	1.0	1.1
1879	1.7	30.8	23.2	.9	5.7	16.6	7.6	2.1	2.7	2.8
1889	2.5	44.8	31.6	1.4	7.5	22.8	13.2	2.2	3.6	7.4
1899	3.2	58.0	33.3	1.7	9.0	22.6	24.7	4.9	7.7	12.1
1909	3.6	66.0	32.4	2.0	9.8	20.6	33.6	5.9	10.0	17.7
1914	3.7	69.1	30.3	2.1	10.1	18.1	38.8	5.5	10.1	23.1
1919	3.8	72.8	24.3	2.2	9.0	13.0	48.5	5.9	12.2	30.5
1924	4.4	89.2	26.5	2.7	11.8	12.0	62.7	6.1	15.9	40.7
1929	3.1	99.0	25.5	3.0	10.9	10.4	73.5	6.8	20.5	46.2

Detail may not add to totals because of rounding.
* Less than 50 million pounds.

Source: Author's estimates using Vial, 1940, and Straus and Bean, 1940.

DEVELOPING A PRICING SYSTEM

During the period of atomistic competition, pricing had posed a relatively straightforward problem of "discovering" the going price and delivering as much milk as customers wanted at that price. Milk that wasn't sold was made into butter in the farm churn. In contrast, pricing in the newer system presented different problems, which reflected the characteristics of milk production and marketing discussed in Chapter 2...

o It was the <u>flow</u> of milk which had to be priced, not any given lot.

o Production of milk was variable from day to day and markedly so seasonally, while demand for fluid milk was comparatively stable over time.

o There were significant costs of producing milk for fluid use which were not involved in producing milk exclusively for manufactured products.

o The costs of transporting fluid milk and its products (cream, butter, etc.) varied widely, with the transportation rate for 100 pounds of fluid milk over a given distance 8 to 10 times as great as for the same amount of milk-equivalent in 40-percent cream and 20 to 30 times as great on the same amount of milk made into butter or cheese.

o The variability of supply in the relatively short run meant that supply could not be tailored precisely to demand, even if the demand were known with precision. A reserve of milk was needed to be able to meet demand on any given day.

o The additional costs of producing milk for fluid use had to be met or no farmer would be willing to meet the sanitary requirements for the fluid milk market.

o Milk used in the different classes of products was in competition with different areas because of the wide differences in transportation costs. Butter was sold in a national market but the market for fluid milk was much more confined geographically.

There were no precedents to be followed in devising a workable method of pricing in such a situation. Consequently, much experimentation occurred and tensions often ran high. Producers were often in conflict with dealers, especially in large markets. Organizing efforts among producers were recurring, in many cases ending in failure in a few years and then starting again some years later. In general, organizations with staying power were not created until just before World War I in large markets and the twenties in many medium-sized markets. Thus, Spencer and Blanford have characterized the period from about 1880 to about 1916 as the era of dealer dominance.

In the beginning, flat pricing was universal. Dealers paid producers one price for all milk, the price being sufficiently above that paid by manufacturing plants to induce producers to incur the expense of meeting sanitary standards for the market in question.

The problem was to limit milk production and receipts to the quantities needed by the dealers, especially in the flush season. Dealers could add producers in the short season and cut them off in the flush, but this created the kind of instability discussed earlier in Chapter 2.

Base-excess plans were developed in an attempt to deal with this problem. Each producer was assigned a base of his deliveries in the short season and was paid the fluid milk price for that quantity year round. Excess quantities were paid for at manufactured product values. Some of the plans were instituted by dealers or their organizations and others negotiated by producer organizations. Base-excess plans were an improvement over flat pricing, i.e., markets where they were in use were somewhat less inherently unstable, but they assumed that all change was seasonal which was hardly the case. Also, they had the fundamental defect from the point-of-view of many producers that neighboring producers often received very different prices for their milk. That hardly seemed like equity.

Classified (or use) pricing was developed to deal with these problems. The highest (Class 1) price was paid for milk in fluid uses, a somewhat lower price for fluid cream, and a still lower price--competitive with the manufacturing milk price--for that milk actually used in manufactured products. Such plans recognized the differences in transportation costs between milk, cream, and manufactured products and the resulting difference in value of milk in different uses. They also recognized the need in the month of lowest milk production for some excess over current fluid milk needs to provide a margin of safety against unusual jumps in demand or sharp drops in supply because of winter storms or other causes. They also recognized the higher costs of producing milk for fluid use occasioned by stringent sanitary regulations governing the care and stabling of cows, types of equipment used, methods of sterilizing equipment, and many other matters.

A classified price plan was first introduced in the Boston market about 1886. The Milk Producers Union negotiated the plan with the large dealers who handled about three-quarters of the milk in the market. In 1901, it was replaced by a base-rating plan (Cassells, 1937, pp. 58 and 67; Whitaker, 1898 and 1905). In 1918, a classified pricing program was reintroduced in the Boston market by the New England Milk Producers Association (Cassells, 1937, p. 58). Thereafter, the adoption of the two-price plan was rapid in other large city markets...in Minneapolis-St. Paul in 1918, in Baltimore in 1919, in Philadelphia in 1920, and in New York and Milwaukee in 1921 (Gaumnitz and Reed, 1937, pp. 31 and 195). Numerous other markets adopted such plans in the twenties and early thirties. In every case, a cooperative negotiated its adoption with the larger dealers.

Many problems arose in the administration of such plans. It was necessary, in order to preserve market stability, that all the available or inspected milk be sold. If some milk producers who held health department shipping permits were turned away without a market for their milk, it would be in their interest to cut their prices and to attempt to find a buyer at the lower price. In some cases, there was literally no alternative market for a displaced producer of fluid milk because manufacturing milk plants unable to meet fluid milk prices had closed down or moved to manufacturing milk territory. If a cooperative could not sell all the milk of its members, it would lose members and bargaining influence in the market.

Pooling was a necessary complement to classified pricing. Producers were paid a weighted average price, reflecting the amount in different uses. The pool could cover the milk sold to each handler but there was wide variation in their Class 1 utilization. The variance among 62 distributors in five Federally regulated markets in the mid-thirties was also typical of earlier periods, ranging from less than 10 percent in Class 1 to over 90 percent.

Percent of total purchases used in Class 1	Number of distributors
0-19	7
20-39	5
40-59	27
60-79	13
80-100	10
Total	62

Source: Gaumnitz and Reed, 1937, p. 30.

Milk production fluctuated widely between spring and late fall, whereas the demand for milk for fluid use was relatively stable. As a consequence, markets experienced a heavy seasonal "surplus" of milk in April, May, and June of each year and were short of milk in October, November, and December. There was a tendency for the milk-shed to enlarge during the short season by addition of new producers. These producers, once in the market, wanted to stay. As new producers were added in the fall, the succeeding flush production surplus was aggravated. Milk distributors complained of having too much milk and used the situation as means of attempting to beat down the the price and the cooperative had the choice of taking in the new producers as members or facing a shrinking market share.

In an effort to level out production over the year, cooperatives developed what came to be known as the base-surplus or base rating plan, which combined features of the base-excess plan with classified pricing. Under this plan, a base was established for each producer at the average level of his shortest production months. During the following year, each producer would be paid each month a base price for his deliveries up to a quantity of milk equal to his base. The remainder of his milk would be paid for at the surplus price. The base price was an average price for all milk sold in fluid form as milk or cream, together with any lower-class milk, up to an amount which equaled the total bases of all the members of the cooperative. A plan could be applied either on the basis of an individual dealer pool or of a marketwide or association-wide pool. The classified price plan was a way of pricing milk to dealers in accordance with its use. The base rating plan was a means for distributing returns for milk in different uses among the producers as an incentive to development of more even production of milk from month to month and season to season. Theoretically, if the total base of all producers were equal to an average Class 1 and Class 2 (cream) sales, a producer who managed to produce exactly his base amount each month would always be paid an average of Class 1 and Class 2 prices with no lower-priced manufacturing milk involved. Actually, the prices received for base milk frequently led producers to make strenuous attempts to increase their own bases each year with the result that the base price itself was diluted by inclusion of lower-class milk.

PUBLIC POLICY CONCERNS

Public policy has been concerned with the milk business for nearly a century. The earliest concerns were with public health aspects of the fluid milk business, since milk-borne diseases were a significant public health problem. Massachusetts passed the first pure milk law in 1856 (Russell, 1976, p. 356). The concern was with fluid milk products, since the bacteria that were the source of the problem were killed or severely disabled in the manufacture of products such as butter and cheese. The fluid milk business was a local one, so local health authorities were responsible.

Early regulation was principally in the large cities, where the need was more acute because of the greater distance, time, and number of agencies involved. Regulation was carried out under the police power of the States, reserved to them by the separation of powers in the Constitution. In most States, this power was delegated, at least in part, to local governments, both counties and municipalities.

Most large cities had sanitary regulations before 1920. In the twenties, the remaining large cities and many smaller municipalities enacted such regulations. The United States Public Health Service had become interested in the problem of milkborne diseases and their sources early in the century and assumed an active role in encouraging enactment of such regulations in the twenties. In 1924, PHS developed a model regulation, the "Standard Milk Ordinance," for voluntary adoption by State and local milk sanitation authorities. To provide uniform interpretation of this ordinance, an accompanying code was published in 1927. The model milk regulation was revised 12 times between 1924 and 1963 (U.S. Public Health Service, 1963, p. 1).

The model ordinance and code encouraged uniformity in the terms of the regulation. By 1932, it had been adopted by 5 States, 86 counties, and 482 municipalities. Adoption of the standard regulation did not guarantee, of course, uniform application or interpretation, but it made such uniformity between jurisdictions possible.

Sanitary regulations served a clear-cut purpose of protecting public health. On this, there was very little disagreement, either as to need or effect. They also made possible the use of sanitary regulations as barriers to movement of milk for economic purposes. One study of the State milk laws--but not the local regulations--concluded that they were primarily to prevent unwholesome milk from consumption channels before 1929, but subsequent legislation was entering primarily to better or maintain the economic status of a particular group in the dairy industry (Rada and Deloach, 1941). While the basis of public health regulation of milk supplies, processing, and distribution is protection of the public from milk-borne diseases, such regulation can have major economic effects. It raises the cost of entry into production of milk for fluid use and its processing and can be used to exclude producers, processors, or both from a given market.

The second major area of public policy concern to arise was in pricing the flow of milk for use in fluid products. Because of the characteristics of milk production and marketing which were discussed in Chapter 2, especially the inherent instability, this has been the major area of concern throughout this century. One key result is that, with raw milk, the pricing system is designed to deal with flow of product. All of the features of the pricing scheme are, and must be, designed to deal with a flow. The characteristics of an individual pail, can, or tankload of milk are of minor importance...they only acquire importance as part of the stream of milk in the system. This fact is highly significant in understanding how and why the milk pricing system developed in the particular form that it took. Without grasping this notion, one can only misunderstand the classified (use) pricing system for milk.

Pricing a continuous flow differs from batch pricing which is typical of agricultural products because it is so inefficient and costly to price each delivery of milk that it is impossible to do so. Prices must be established to cover all deliveries during a given period, typically a month although longer or shorter periods are also possible. The market is no longer cleared with each transaction but only over the longer period.

The fact that 99.44 percent of the milk is priced on a flow basis does not mean that there are no spot markets. Such markets have existed for nearly a hundred years but they deal only with the marginal lots of milk needed to meet a dealer's commitments.

The third area of public policy concern was income support for dairy farmers. The Agricultural Marketing Act of 1929 established the Federal Farm Board "to promote the effective merchandising of agricultural commodities." The Board made loans from a revolving fund to cooperatives and stabilization corporations, with the losses resulting from stabilization activities borne by the revolving fund. Some loans were made to dairy marketing cooperatives to hold butter off the market "until prices improved." In May 1933, the powers of the Board were consolidated with those of other credit agencies to form the Farm Credit Administration (Rojko, 1957, p. 150).

Concern for income support came to a head during the Depression. The approach to the problem during the thirties was largely through direct participation of Federal and State Governments in the pricing of milk for fluid use. Some efforts to raise producers' incomes by purchasing manufactured dairy products were also made. The price support program for dairy products acquired its basic form during World War II.

The concerns with price and income problems have become inextricably linked. Given the characteristics of milk production and marketing previously discussed, it could hardly be otherwise. These are the public policy problems which are the concerns of this book.

The final area of public concern was over the policy toward agricultural marketing cooperatives. They had been treated by the antitrust community as illegal conspiracies in restraint of trade until 1922 when passage of the Capper-Volstead Act confirmed their partial immunity from the antitrust laws (see Manchester, 1982, and Guth, 1982). The earliest public involvement extended only to mild encouragement of cooperatives through educational efforts. Doubts about the legality of the existence of farmer cooperative marketing associations under the Sherman Act of 1890 had not been cleared up by specific provisions in the Clayton Act of 1914, so farmer cooperatives were subject to legal action challenging their existence. The 1922 Capper-Volstead Act clarified the legal status of cooperative marketing associations. The Cooperative Marketing Act of 1926 specifically included farm supply cooperatives and established a program within the Department of Agriculture to assist cooperatives in their activities.

IN SUMMARY

In two centuries ending in 1933, the U.S. dairy industry had changed from a relatively primitive farm operation where most cows produced milk only in the Summer and nearly all milk was used on the farm, much of it for butter and cheese, to a mostly small-scale commercial farming industry. There were still several million farms with one or two cows producing for home use but a commercial dairy industry producing both for fluid milk and for manufactured dairy products had developed in the late nineteenth century.

Up until about 1880, the competitive picture in fluid milk markets was largely one of atomistic competition. From then until World War I, large dealers dominated. Cooperatives became important in negotiating with dealers during and after the War.

In this period, the pricing problem in fluid milk markets took shape and the classified (use) pricing system was developed to deal with it. Public concerns were focused largely on sanitary regulation, with some attention to encouraging the growth and development of cooperatives.

Part 2

Industry Structure

4
Dairy Production and Marketing in the Early Thirties

In the broadest terms, the conditions which led to the Federal dairy programs and many State programs intended to deal with the same problems were the result of the Great Depression. For the general economy, the depression began with a bang with the Great Stock Market Crash of 1929. But the depression for most of agriculture had begun in the early twenties. Dairy farmers had fared better than farmers in general, especially those raising crops, during the twenties. Cash receipts from farm marketings of dairy products (milk, cream and farm butter) rose from $1.4 billion in 1924 to $1.8 billion in 1929, 31 percent, while those of the rest of agriculture rose only 7 percent.

Prices for milk sold to fluid milk distributors (not all of which was used in fluid products) were above parity throughout the twenties, i.e., they were more favorable than the 1910-14 relationship between prices received and prices paid by farmers. Prices for all milk sold at wholesale (i.e., to manufacturing plants as well as fluid milk dealers) were below parity from 1924 to 1929, standing at 93 percent of parity in 1929 (Black, 1935, p. 463). This is parity as then defined and calculated. Using the "modernized" parity of the 1949 Agricultural Act, prices were closer to parity--101 percent in 1927 and 97 percent in 1929.

The marketing arrangements of producer cooperatives were partly responsible for the differing performance of dairy farming and the rest of agriculture in the twenties, although other factors were more important. The somewhat better--i.e., not-as-bad--position of dairy farming in 1932 also led to measures to assist dairy farmers being almost an afterthought in the Hundred Days of the Roosevelt Administration when the first Agricultural Adjustment Act was conceived and passed (Black, 1935; Nourse, 1935).

The Depression brought a sharp drop in milk prices. The average price of milk sold to distributors dropped 31 percent, from $2.81 per hundredweight in 1929 to $1.72 in 1932. This was caused partly by the dramatic decline in purchasing power due to widespread unemployment and sharply-reduced incomes of those still working. Total payrolls dropped 57 percent between 1929 and 1932. It was also due in part to chaotic conditions in the markets where milk producers sold.

In late 1932 and early 1933, producer prices for fluid milk broke almost completely in most markets. For a short time in some markets, organized producers actually received net prices lower than butterfat prices plus the value of skim milk, i.e., less than manufactured product value (Black, 1935, p. 82).

Feelings of desperation on the part of dairy farmers led to milk strikes in a number of markets, including one in New York in August 1933. Even though State control had gone into effect in late May and prices to farmers for Class 1 milk had been raised 35 cents per hundredweight on July 21 because of a severe drought, farmers had not yet received the higher prices and went on strike (Spencer, 1941, p. 4). There were also strikes in Chicago, Wisconsin, and Iowa.

MILK PRODUCTION AND FARM SALE

In the twenties and thirties, a substantial proportion of American farmers still kept one or more milk cows to provide milk for family consumption and, in part, for livestock feed. Only 62 percent of farms with cows sold milk, cream, or butter. And only 19 percent sold whole milk--either for fluid milk use or for manufactured products. Commercial dairy farms--those where dairy products accounted for 40 percent or more of receipts--were only 13 percent of the total.

With the Depression, even more farms added one or more cows and the number of farms reporting milk cows increased 11 percent. Over three-fourths of all farms had one or more milk cows in 1934.

Milk production on farms (an estimated 2.8 billion pounds per year was still produced by nonfarmers largely for use in their own households) increased a bit during the Depression, but the increase was not reflected in sales off the farm (table 2). About a quarter of all farm milk was used on the farm, a substantial part for farm-churned butter (some of which was sold) and about a third was skimmed on the farm and the butterfat sold to creameries (butter manufacturing plants).

Retailing of milk by farmers increased, as other outlets became increasingly difficult to find. Retailing also provided an outlet for excess farm labor, especially family members who could not find other jobs.

Half the herds in 1929 had one or two cows...78 percent less than 10 cows (U.S. Bu. of Agric. Econ., 1950, p. 2034). Over half the cows were in herds of 6-20 cows and they produced 56 percent of all milk in 1932 (table 3). Milk production per cow rose with size of herd. The farms with only a few cows used most of it at home or sold it as farm-separated cream, so those with 1-5 cows produced 28 percent of the milk but sold only 7 percent of the whole milk.

The small herds were concentrated in the South, with half of the cows in that region in herds of 1-3 cows. Utilization reflected the differences in herd sizes. While a quarter of all milk was used on farms nationally in 1934, the proportion in the South Atlantic region was 62 percent. Two-thirds of the milk produced in the North Atlantic region was sold as whole milk, while a similar proportion of the milk in the West North Central region was separated on the farm and sold as cream (Gaumnitz and Reed, 1937, p. 9).

Table 2--Milk production, use, and marketings
by farmers, 1929-35

Year	Milk produced	Milk used on farms where produced	Milk marketed
	- - - - - - - - - - million pounds - - - - - - - - - -		
1929	98,988	24,990	73,998
1930	100,158	24,840	75,318
1931	103,029	25,971	77,058
1932	103,810	27,186	76,624
1933	104,762	27.460	77,302
1934	101,621	26,804	74,817
1935	101,205	26,017	75,188

Source: Crop Reporting Board.

Table 3--Size distribution of farms with dairy cows, 1929-32

Number of cows per herd	1929	1932		
	Percent of cows	Percent of cows	Percent of milk produced	Percent of whole milk sold
1	6.5	5.8	5.0	0.4
2-3	14.9	13.6	10.9	2.2
4-5	14.3	13.5	11.6	4.8
6-10	28.2	28.0	27.5	22.5
11-20	23.8	25.8	28.5	37.9
21-30	6.5	7.5	9.1	16.5
31-50	3.4	3.5	4.3	8.8
More than 50	2.4	2.3	3.1	6.9

Source: Gaumnitz and Reed, 1937, pp. 6 and 8, from Crop Reporting
Board.

Production per cow declined steadily from a peak of 4,579 pounds in 1929 to 4,033 pounds in 1934, primarily because of the effects of drought on pasture conditions. Other factors were less grain feeding and lower culling rates. It is quite probable that there was also an increase in the proportion of dual-purpose cows (principally Shorthorns) which were milked. With their substantially lower output per cow, the average was undoubtedly pulled down. A 1936 survey found that 26 percent of the Corn Belt (Ohio, Indiana, Illinois, and Iowa) farms with cows had Shorthorns and 29 percent of the Northern Plains (North and South Dakota, Nebraska, and Kansas) farms with Shorthorns. In contrast, none of the Northeastern farms had Shorthorns and only 3 percent of those in the Lake States (Elwood et al, 1941, p. 22).

Dairy farming in the thirties was a small-scale, hand-labor operation. A survey by the Works Progress Administration and the Bureau of Agricultural Economics found little mechanization and that only on large farms (table 4). Silage feeding had become fairly general, especially in the Lake States, but pastures and hayfields in the Northeast were without alfalfa or clover on more than three-fourths of the farms.

Half the farmers in the Lake States and nearly a third in the Northeast were still hauling their own milk or cream to the plant and some were still using horses to do so. These figures are dominated by the small farms which were so numerous. Larger, specialized dairy farms were much more likely to be mechanized. Milking machines were found on only 3 percent of the Champlain Valley farms with less than 14 cows in 1932, compared to 63 percent of farms with 38 or more cows (Williams, 1941, p. 52). Milk production was about as labor-incentive as it had been for thirty years. Hours of labor per cow were little different than they had been before World War I...

1909-13......	135 hours per cow
1917-21......	138
1927-31......	139
1932-36......	140
1937-40......	139

But milk production per cow was increasing throughout the twenties, so that hours of labor per thousand pounds of milk decreased somewhat...

1909-13......	35.5 hours per thousand
1917-21......	36.5 pounds of milk
1927-31......	30.7
1932-36......	33.0
1937-40......	31.0

By 1932, incomes of dairy farmers had dropped to nominal levels, if they were not negative. The Champlain Valley of Vermont was a profitable dairy area in the twenties but barely provided subsistence in 1932 (Williams, 1941, p. 32).

Table 4--Characteristics and practices of farms with milk cows,
Northeast and Lake States, 1936

Characteristics and practices	Northeast	Lake States
Average number of milk cows	17.0	17.0
Percent of farms feeding silage	62.0	85.0
Silage feeding period (months)	6.9	8.1
Percent of farms feeding:		
Legume hay (any kind)	22.0	71.0
Alfalfa	13.0	62.0
Mechanization:		
Percent of farms with water bowls	44.0	58.0
Removing manure by--		
Hand	44.0	49.0
Carrier	15.0	38.0
Manure spreader or wagon	23.0	13.0
Scuttle	18.0	0.0
Feeding method:		
Hand	89.0	86.0
Truck	7.0	8.0
Carrier	4.0	6.0
Milking machine	14.0	29.0
Hand milking	86.0	71.0
Cooling milk or cream:		
Any method	99.0	94.0
Water, ice, or both	84.0	93.0
Aerator	2.0	1.0
Electricity	13.0	*
Separating cream on the farm	3.0	38.0
Percent of farms hauling own milk or cream	31.0	52.0
Method of hauling (pct. of farms):		
Automobile	24.0	52.0
Truck	60.0	44.0
Horse	16.0	4.0
Average number of trips per week	6.7	5.6

*Less than 0.5 percent.

Source: Elwood et at, 1941.

	1926	1932
	- - dollars per farm - -	
Income from farming	1,407	104
Food, fuel, and housing	509	386
Total	1,916	490

Farmers with borrowed capital of any magnitude would not have had enough cash flow to cover interest payments.

MARKETING BULK MILK

The universal method of transport for milk and cream from farm to receiving station or processing plant was in 40-quart cans throughout the twenties and thirties. By the thirties, trucks carried almost all milk which was picked up on regular routes, although the numerous small producers—especially cream-shippers—often carried their own milk to the plant on the way to town. Milk for fluid use was picked up every day, with that for manufactured products sometimes less often.

Most assembly routes were organized by the processor, although the drivers could be employees of the processor, independent contractors, or otherwise engaged.

A 1929 study found 46 percent of New York State farmers transporting their own milk to the country plant every day, 44 percent shipping by commercial truck, and the remaining 10 percent cooperating with one or more neighbors (Spencer, 1929, p. 12). The average distance from farm to plant was 7 miles. In Connecticut in 1936, 49 percent of the trucks were owned by dealers, 20 percent by independent commercial haulers, and 31 percent by farmers (Hammerberg and Sullivan, 1942, p. 5). Dealers hauled 60 percent of the milk, haulers 36 percent, and producers only 14 percent.

Hauling charges came down during the twenties and thirties, as trucks and roads improved and, then, prices dropped during the Depression (Hammerberg and Sullivan, 1942, p. 14).

Average Hauling Rates in Connecticut
Cents Per Hundredweight

Year	Rate
1922	46
1927	41
1932	37
1933	35
1934	31
1935	29
1936	27
1937	25
1938	25
1939	25
1940	24
1941	14

The Connecticut rates were very similar to those in four other markets in 1937 (Louisville, Chattanooga, Indianapolis, and Fort Wayne) which ranged from 22.7 to 25.0 cents (Roadhouse and Henderson, 1950, p. 180).

Most large fluid milk markets were self-sufficient. In other words, they had sufficient supplies of fluid-grade milk in the milkshed contiguous to the market to meet the needs for fluid milk products, including a reserve against seasonal, week-to-week, and day-to-day variations in supply and demand (Herrmann, 1950, p. 6).

The size of milksheds varied, of course, depending on the size of the market, density of production, and the existence of other markets nearby. Milwaukee, in the midst of a heavy milk production area, obtained most of its milk in 1937 within a radius of 10-20 miles and 99 percent from 40 miles or less (U.S. Agric. Adj. Admin., 1937b, p. 130).

In the Northeast, the New York and Boston milksheds extended several hundred miles, with much milk traveling more than 250 miles. Philadelphia's receipts were larger than Boston's, but nearby production was heavy and it obtained most of its milk within 60 miles of the city (Spencer, 1938, p. 12).

Los Angeles obtained most of its milk from within its extensive city limits. Production was almost entirely drylot dairying, i.e., feed and forage were shipped in. San Francisco and Oakland, surrounded by water and mountains, obtained most of their supply from coastal counties north of San Francisco and from the San Joaquin Valley (Beal, 1944, p. 6).

Receiving Stations

The first movement of milk from the country to the city in the nineteenth century by producers located within a few miles of a railroad station involved the producer hauling milk in his own cans to the railroad (and later the trolley) station and often loading the milk cans onto the train himself. Soon, shipping stations arose at points where farmers brought their milk, ranging from an open platform beside the tracks to a solidly built shed. On the open platform, milk was often exposed to the sun and heat for some time before the train arrived. About 1885, country receiving stations began to be built by the railroads, producers' organizations, and city distributors at points from 30 to eventually as much as 400 miles from the city. Those owned by the railroads were eventually sold to distributors or producer organizations. Some of these stations also skimmed, pasteurized, and manufactured milk, in addition to weighing, testing, and shipping it to market.

The advent of the motortruck and its spread after World War I gradually replaced many receiving stations. Peak numbers of receiving stations were probably reached in many markets between 1917 and 1923. As trucks replaced wagons in hauling to the receiving station, many fewer stations were needed to handle the same volume of milk and direct hauling of milk from the farm to the city distributor's plant replaced many close-in receiving stations.

By 1934, there were no rail receipts of milk in 17 of 28 large markets and very little in another 6 markets (table 5). Only in Boston and New York did rail account for a majority of receipts. In these truck-only markets, there were no receiving stations in eight markets and three or less in another 10 markets. In the smaller markets not included in this study, supply areas were more compact and nearly all receipts were direct-hauled by truck from the farm to the distributor's plant.

In the large markets where the milksheds reached out several hundred miles--New York, Boston, and Chicago are the main cases--transportation of milk from country plants to city plants was very largely by rail in tank cars. In Boston during 1936, trucks were used exclusively within 80 miles of the market, but accounted for less than 20 percent of shipments up to 200 miles and almost none beyond that point (Pollard, 1938, p. 20). Three tank cars operated in the Boston market in 1907--probably the first use of such cars (Kelly and Clement, 1931, p. 211). By 1938-39, about 65 tank cars were operating and they hauled 86 percent of the milk moved by rail (Sonley, 1940, p. 8).

In contrast, Philadelphia with its relatively compact supply area received 71 percent of its milk by truck in 1935 and 86 percent in 1941 (U.S. Bu. Agric. Econ., 1942).

The relatively few long-distance shipments of milk required to meet occasional seasonal shortages in a market were by rail in 40-quart cans or occasionally in tank cars. However, tank cars were usually owned or leased by regular shippers, so they were not available for irregular shipments.

Nearly all large and most middle-sized cities had adopted sanitary regulations for milk by the early thirties, as had a number of states. These regulations had reduced the incidence of milk-borne diseases sharply. They had also effectively differentiated milk which met the standards from that which was eligible only for use in manufactured products. These regulations also provided a ready-made opportunity for local interests--both producers and processors--to put them to use to isolate the local market from outside competition. In many cases, the local health department joined in enthusiastically. The most effective tools were refusals to allow into the market any milk not produced on farms inspected by the local health department and refusal to allow packaged milk into the market from a plant outside the city limits or a radius of, say, 10 miles.

Role of Cooperatives

While the history of milk marketing cooperatives goes well back into the nineteenth century, most of the pre-depression development began just before World War I. Their role in manufactured dairy products relates to processing and will be discussed in the next section. In fluid milk marketing, cooperatives were organized in many large and middle-sized markets by 1933. In 1927, 159 cooperatives marketed 11 billion pounds of bulk whole milk (Metzger, 1930). Most of these were bargaining associations confined to a single fluid milk market. In the largest markets, cooperatives such as the

Table 5--Receiving stations and transportation methods, 28 fluid milk
markets, 1916 and 1934

Market	1916 Receipts by railroad	1934 Receipts from re-ceiving stations Railroad	Truck	Direct hauled by truck	Receiving stations
		----------Percent-------------			Number
Detroit, Mich.	74	0	97	3	54
New York, N.Y.	98	63	33	4	550
Boston, Mass.	90	84	4	12	95
Philadelphia, Pa.	87	31	36	33	84
Chicago, Ill.	100	33	33	34	72
Pittsburgh, Pa.	*	3	52	45	24
Baltimore, Md.	75	3	49	48	16
New Orleans, La.	65	33	17	50	9
St. Louis, Mo.	75	0	49	51	29
Kansas City, Mo.	*	0	17	83	2
San Francisco, Ca.	60	5	10	85	3
Cleveland, O.	95	0	10	90	9
Seattle, Wash.	*	0	10	90	3
Minneapolis-St. Paul, Minn.	*	0	10	90	2
Washington, D.C.	50	1/	9	91	2
Los Angeles, Ca.	*	1/	5	95	3
Milwaukee, Wis.	60	2	3	95	2
Louisville, Ky.	*	0	4	96	1
Columbus, O.	*	0	2	98	1
Spokane, Wash.	*	0	2	98	1
Cincinnati, O.	75	0	0	100	0
Dayton, O.	*	0	0	100	0
Des Moines, Io.	*	0	0	100	0
Indianapolis, Ind.	73	0	0	100	0
Omaha, Nebr.	*	0	0	100	0
Portland, Oreg.	*	0	0	100	0
Richmond, Va.	*	0	0	100	0
Sioux City, Io.	*	0	0	100	0

* Not available.
1/ Less than 0.5 percent.

Source: Scanlan, 1937, p. 26.

Dairymen's League Cooperative Association of New York operated country plants and receiving stations, serving secondary as well as primary markets. The League had 273 country plants in 1930 and, despite many closings, 117 in 1936 (U.S. Fed. Trade Comm., 1936, p. 39).

Cooperative shares of total market receipts of milk in 1932 in six major Northeastern markets ranged from 42 to 97 percent (Stitts et al, 1933, p. 15):

Boston	51 percent
New York	42
Philadelphia	69
Baltimore	97
Washington	94
Richmond	71
Six markets	51

In 1934-35, 110 cooperatives were bottling and distributing milk. They had 70,000 members and sales of $90 million, about 5 per cent of fluid milk sales of all commercial processors. The 87 milk bargaining associations had 172,000 members and had total sales of $149 million, about half of all fluid grade milk. (Elsworth, 1936, pp. 33, 41).

PROCESSING MANUFACTURED PRODUCTS

Processing of manufactured dairy products has always varied widely from one sub-industry to another, so they will be discussed separately.

Butter

Butter was and still is the residual use of whole milk and the only use of farm-separated cream. Some butter was made in fluid milk areas from seasonal surplus milk. Country plants in these markets often had a churn for this purpose.

In manufacturing milk areas, local creamery butter was made from more or less sweet cream or whole milk delivered every day or two to nearby small creameries. It was generally of good quality but poorly standardized as to color, flavor, and other characteristics. Large sales organizations, especially Land O'Lakes Creameries, Inc., had made substantial progress in improving quality and especially in sorting the butter. Centralizer creamery butter was made mostly from sour cream shipped or hauled longer distances at less frequent intervals. It was more uniform in quality but made of poorer raw materials (Black, 1935, p. 39). Considerable butter was still made on the farm from hand-skimmed or machine-separated cream. In 1932, 1 percent of all butter sold came from farm churns, in addition to 93 million pounds (4 percent of total consumption) made and consumed on farms.

There were 3,527 plants whose principal product was butter in 1929 (U.S. Bu. Agric. Econ., 1950, p. 1991). Cooperatives made 36 percent of the butter in 1933 and 1934 (Elsworth, 1936) and 39 percent in 1936 (Hirsch, 1951). Most of the 1,385 cooperatives with butter as the major product were local creameries. Cooperative sales agencies sold 27 percent of the butter produced by cooperatives in 1934 (Elsworth, 1936).

Large companies--both dairy firms and meat packers--were early entrants into the assembly and distribution of manufactured dairy products. In 1918, four large meat packers sold 19 percent of total butter, very little of which was produced in their own plants (U.S. Fed. Trade Comm., 1921). Three large dairy companies accounted for another 13 percent. The two largest dairy companies produced 61 percent of the butter that they sold, 6 percent of total creamery butter production. Between 1918 and 1934, the share of Swift and Armour, the two largest meat packers, in butter sales declined from 19 to 15 percent. Eight other meat packers sold 4 percent of the butter in 1934 (U.S. Fed. Trade Comm., 1937, pp. 228, 247). The three largest dairy companies sold 16 percent of creamery butter and nine others 9 percent.

Cheese

Most natural cheese--both American and foreign types such as Swiss and Italian--was made in small cheese factories typically with 15-35 farmer patrons, half of them living within a half-mile of the factory. There were 2,758 cheese factories in 1929, most of them in Wisconsin (U.S. Bu. Agric. Econ., 1950, p. 1991). In 1936, the 562 cooperatives making cheese accounted for 25 percent of production (Trelogan and Hyre, 1939), down from 32 percent in 1926.

In 1934, 41 percent of natural cheese (excluding soft cheeses) was sold (not produced) by two large dairy companies and another 3 percent by two other large dairy firms (U.S. Fed. Trade Comm., 1937, pp. 228, 247). Meat packers were also important in cheese assembly and distribution. Four large meat packers handled 41 percent of natural cheese sales and two others 4 percent.

Local cheese factories sold to wholesale cheese buyers (assemblers). Many cooperatives sold through cooperative sales agencies. About 30 percent of American cheese was made into processed cheese and cheese food, mostly by the large cheese companies such as Kraft and Borden.

Evaporated and Condensed Milk

Canned evaporated and condensed milk was made largely by a few national dairy organizations who sold under national brands. Seven large dairy firms sold 68 percent of the canned milk in 1918--the four largest 54 percent (U.S. Fed. Trade Comm., 1921). Two meat packers accounted for another 9 percent. By 1934, the four largest dairy companies' share had declined to 49 percent and one other had 1 percent (U.S. Fed. Trade Comm., 1937, pp. 228, 247).

A few retail grocery chains also manufactured canned evaporated milk for sale under their own brands. A&P accounted for 12 percent of canned milk in 1934. Three meat packers produced 7.5 percent (U.S. Fed. Trade Comm., 1937, p. 247).

Independent condenseries made canned evaporated and condensed milk for sale through grocery channels, some of it under the buyer's label. They also made bulk condensed products, selling them either through brokers or under contract with manufacturers of food products using concentrated milk. A few milk distributor organizations operated condenseries as sideline enterprises to utilize a part of the surplus from fluid milk operations, as did a few fluid milk cooperatives which handled their own surplus (Black, 1935, pp. 40-41). Cooperatives accounted for only 3 percent of the production of evaporated and condensed milk in 1936 (Trelogan and Hyre, 1936).

Ice Cream

Several diverse elements made up the ice cream industry in the early thirties. There was a well-developed specialized ice cream manufacturing and distribution industry with a strong element of national dairy companies. A separate trade organization of ice cream manufacturers dated back to 1900. Many fluid milk processors also operated ice cream plants, sometimes as a part of a single plant which also packaged fluid milk products. Drug stores, candy stores, and restaurants often made their own ice cream. Specialized ice cream stores began to be established in the thirties; some made their own ice cream and others were owned or franchised by ice cream manufacturers. Cooperatives played a minor role in ice cream manufacture, accounting for less than 1 percent of the total in 1936 (Trelogan and Hyre, 1936).

Cottage Cheese and Soft Cheeses

The manufacture and distribution of cottage cheese and soft cheeses (cream cheese, Neufchatel, etc.) were an adjunct of the fluid milk business. Cottage cheese was almost entirely made in fluid milk plants and was distributed on milk distribution routes.

FLUID MILK PROCESSING

In 1932-33, the organization of fluid milk processing and distribution varied widely between town and city. In villages and small towns, the business was very largely in the hands of producer distributors--farmers who produced, bottled, and distributed milk. In larger towns and smaller cities, the business was typically shared between producer-distributors and one or more small commercial processors. Producer-distributors were more important in the South and West, with 60 percent of the volume in Atlanta, 40 percent in Birmingham, 50 percent in Kansas City, Missouri, and 40 percent in Omaha. In contrast, they had 1 percent in Philadelphia and New York City and none in Cleveland. Ordinances requiring pasteurization of milk effectively put most producer-distributors out of business. In the larger cities, producer-distributors still played a part but commercial distributors were definitely dominant.

Nationally, producer-distributors retailed 5.3 billion pounds of fluid milk products in 1929 (25 percent of all fluid milk products sold), 5.6 billion pounds in 1932 (26 percent), and 5.7 billion pounds in 1935 (26 percent of the total). There are no firm statistics on the number of producer-distributors in this period, but they must have numbered more than 35,000. Numbers undoubtedly increased along with volume as the depression deepened and farmers looked for ways to improve incomes and utilize family labor.

Technology was simple for the producer-distributor. He had to put the milk or cream into a bottle (quarts and pints for milk; half-pints for cream), refrigerate it, and deliver it to consumers. The smallest operators probably used a funnel and pitcher to fill bottles. Somewhat larger operators purchased a simple bottle filler and capper. Refrigeration was spring or well water or ice. Most any vehicle would do for distribution, refrigeration not being generally used.

The number of commercial processors of fluid milk increased during the Depression years from perhaps 8,000 or fewer in 1929 to about 8,900 in 1934 and 9,500-9,600 in 1940. In larger cities such as Boston and Chicago, there were well over 100 dealers in the thirties and, in middle-sized cities such as St. Louis, 50 or more (table 6). Typically, city markets consisted of a small number of large plants, many of them operated by national and regional firms. Then, there was a larger number of middle-sized plants operated largely as family businesses (though they might be incorporated). A large "competitive fringe" of small plants including producer-distributors completed the complement of firms in the local market.

In 19 cities for which data are available from a variety of sources for a period in the thirties, the share of the four largest firms in the market ranged from 33 percent to 81 percent, with larger shares more common in the smaller markets (table 7).

A survey of health officers in 71 cities with population of 100,000 or more (out of a total of 93 queried) in 1930-31 found an average of 82 distributors per market in middle-sized cities and 102 in large cities (table 8)--this includes all distributors, whether or not they were bottling milk. The share of those selling raw milk only--mostly producer-distributors--was low in Northeastern middle-sized cities and in large cities, while it was high in the South and West. The share of the three largest firms was roughly 50 percent in the middle-sized markets and 63 percent in the 12 large markets (Spencer, 1932).

These figures inflated to all 93 cities with populations of 100,000 or more imply that there were about 7,900 distributors just in these cities, which had 39 percent of the nonfarm population in 1930. About 3,300 of these distributors were handling raw milk only, mostly producer-distributors. Reasonable estimates for the remainder of the cities and towns would indicate distributor numbers of these magnitudes in 1930--

City size	Number of distributors	Total volume (mil. lbs.)
Over 500,000	1,326	7,800
100,000-500,000	6,560	5,000
All other	25,000-30,000	9,600
Total	32,900-37,900	23,000

Table 6—Changes in numbers of dealers in selected markets, 1930-1942

Market

Year	Boston 1/			Chicago 4/			Omaha 4/			St. Louis 4/		
	No. at start of year 2/	No. of new firms 3/	No. of firms failing 3/	No. at start of year	No. of new firms	No. of firms failing	No. at start of year	No. of new firms	No. of firms failing	No. at start of year	No. of new firms	No. of firms failing
1930	--	--	--	135	9	7	--	--	--	--	--	--
1931	--	--	--	137	11	8	--	--	--	--	--	--
1932	--	--	--	140	5	7	--	--	--	--	--	--
1933	--	--	--	138	11	3	--	--	--	--	--	--
1934	--	--	--	146	7	10	14	0	2	50	1	4
1935	--	--	--	143	6	11	12	0	0	47	0	3
1936	--	--	--	138	1	9	12	2	1	44	1	1
1937	123	11	21	130	4	3	13	1	2	44	2	1
1938	113	24	17	131	0	2	12	1	0	45	0	2
1939	120	34	50	129	6	3	13	1	2	43	0	2
1940	104	6	18	132	10	6	12	--	--	41	2	2
1941	92	--	--	136	0	2	--	--	--	41	0	4
1942	--	--	--	--	--	--	--	--	--	37	--	--
Average	110	19	26	133	6	6	12.6	0.8	1.2	43.6	0.7	2.1
Percent new or failing		17	24		4	4		6	8		2	5

1/ Excluding producer-dealers and dealers who buy milk only from other dealers.
2/ August 1.
3/ New firms and failures includes net changes in numbers of firms which were in business, but were reclassified from, or into, the group of dealers buying from other dealers or the group of producer-distributors.
4/ Dealers operating pasteurizing and bottling plants within the city.
Source: Data for this table were assembled from market administrator's records, health department data and other records. (Herrmann and Welden, 1942).

Table 7--Market shares in 19 markets in the 1930s

Size and name of market	Year	Number of large firms	Market share of large firms
			Percent
Small markets:			
Sioux City, Io.	1937	4	72.5
Quad Cities, Io.	1935	6	46.7
Richmond, Va.	1934-5	3	89.9
Madison, Wis.	1937	3	91.0
Medium markets:			
Worcester, Mass.	1935	4	33.1
San Diego, Ca.	1935	3	63.0
Large markets:			
Louisville, Ky.	1936	4	50.9
Phoeniz, Ariz.	1935	3	84.4
Providence, R.I.	1934	3	42.1
St. Louis, Mo.	1936	4	68.7
Milwaukee, Wis.	1936	4	72.8
Minneapolis, Minn.	1930	4	81.4
Minneapolis, Minn.	1934	4	70.9
Baltimore, Md.	1938	4	74.6
Very large markets:			
Detroit, Mich.	1935	3	52.9
Chicago, Ill.	1935	4	58.8
Philadelphia, Pa.	1936	4	65.8
Boston, Mass.	1934	3	63.2
San Francisco, Ca.	1937	4	67.0
New York, N.Y.	1939	4	54.3

The definition of large firms varies with the source--consistency is beyond hope.

Table 8--Milk distribution in cities over 100,000 population,
1930-31

Item	Cities from 100,000 to 500,000 population					Cities over 500,000 population
	Northeast	Midwest	South	West	Total	
Number of cities	18	14	9	18	59	12
Number of distrib- utors per city	63	80	58	120	82	102
Percent of distrib- utors selling raw milk only	13	41	50	69	49	9
Percent of distrib- utors selling more than 1,500 quarts daily	13	12	12	5	9	45
Percent of total sales by 3 largest firms	48	47	52	57	51	63
Percent change in number of distrib- utors in past 5 years	-7	+18	+10	+29	+16	-13
Average sales volume (quarts per day)	1,265	1,115	1,401	604	976	7,491

Source: Spencer, 1932, p. 1707.

The technology employed by commercial milk bottling plants was relatively simple, although substantially more complex and expensive than that utilized by most producer-distributors. Nearly all milk went into quart or pint bottles, paper containers having been intro- duced in only 2 of 158 markets accounting for 8 percent of the milk distributed by commercial processors. (For more detail on these statistics, see Chapter 12.) Half gallon and gallon containers were not yet in use. Pasteurization was required by health regulations in most large city markets, although not in many smaller towns. Fluid milk product lines typically included--

o Pasteurized creamline milk of minimum butterfat content.
o Pasteurized creamline milk with higher-than-minimum butter- fat--"premium" milk.
o Heavy cream.
o Light cream.
o Buttermilk.
o Chocolate milk (some plants).
o Skim milk (less common).

With this relatively simple product and package line, economies of size in plant operations were largely achieved by middle-sized plants. Many smaller plants undoubtedly offset these economies of size by employing nonunion labor at lower wage rates.

In 1934, three large national dairy companies packaged 4.7 billion pounds of fluid milk products, 32 percent of the total of commercial milk processors (U.S. Fed. Trade Comm., 1937, p. 228). Six regional firms processed 5 percent of the total. Other multi-unit companies operating in several markets probably accounted for a substantial share of all business, perhaps 15-20 percent.

Grocery chain operation of milk plants was in its infancy. Safeway had opened its first plant in Oakland in 1928 and a second in Los Angeles in 1932.

Cooperatives were insignificant in packaging milk in city markets. Only 75 cooperatives had such plants and these were concentrated in small cities (Herrman and Welden, 1941). Some had been organized by producer-distributors when pasteurization became mandatory to do the processing for all members, who continued to produce and distribute independently. The one exception to the small size of cooperatives packaging fluid milk products in the thirties was the Dairymen's League Cooperative Association of New York. While only 15-16 percent of its volume went through its own packaging plants, the League accounted for a high proportion of cooperative packaged milk volume (U.S. Fed. Trade Comm., 1936, p. 38).

MARKETING PACKAGED MILK

As the Depression deepened, it seemed as though everybody went into milk distribution--producer-distributors, cut-rate stores, milk depots, peddlers, and roadside stands. We have already noticed the increase in volume of producer-distributors.

Home delivery every day of the week was the dominant mode of business. In total, nearly three-fourths of all milk was delivered to the doorstep--higher in small towns and cities and lower in very large cities, but with much variation. In the smaller towns and villages of the South, nearly all of the milk was distributed by producer-dealers, many of whom made deliveries both morning and afternoon (Kelly and Clement, 1931, p. 387). When most customers acquired refrigerators, there was no longer any reason for twice-a-day delivery. In 1933-34 in 17 markets for which figures are available, home delivery shares ranged from 57 percent in Pittsburgh to 89 percent in Philadelphia and Binghamton, New York (Black, 1935, p. 45). Hotel, restaurant and institutional sales accounted for about 13 percent of total sales, depending in part on city size (table 9).

Distribution areas for large plants covered all or most of the city and its suburbs, with distribution from the processing plant to depots in various parts of the city. Smaller plants generally concentrated distribution in an area near the plant.

Home delivery sales were approximately equal from Monday through Friday, dropped a bit on Saturday, and rose to a high point about 4 percent above the average for the week on Sunday. Sales to retail stores followed an opposite pattern, constant from Monday to Friday, then rising nearly 30 percent on Saturday and dropping off to about half of the Monday-Friday average on Sunday (Mumford, 1934, pp.

Table 9--Marketing channels for fluid milk products, 1929

Outlet	Processors	Subdealers	Producer-distributors	Total
	- - - - - - - - -Percent- - - - - - - - - -			
Home delivered	51.0	4.6	17.7	73.3
Plant and farm sales	3.6	0.0	2.3	5.9
Stores:				
Grocery stores	3.6	.5	1.5	5.6
Dairy stores	2.0	0.0	0.0	2.0
Other stores	.2	*	.1	.3
All stores	5.8	.5	1.6	7.9
Institutions, hotels, restaurants	11.9	.2	.8	12.9
Total	72.3	5.3	22.4	100.0

*Less than .05 percent.

2036-7). These figures are for New York City--blue laws kept retail stores closed on Sundays in many markets.

The behavior of retail milk prices (home delivered) between about 1909 and the early thirties hardly indicates that conditions approaching perfect competition existed. Indeed, the behavior of prices strongly suggests Gardner Means' classic description of administered prices...unchanged for long periods of time. In 13 cities where the Bureau of Labor Statistics reported retail home delivery prices during all or most of the period from 1907 to 1933, retail prices were unchanged for a year or more at least half of the time in 6 cities and more than 30 percent of the time in 11 out of 13 cities (Gaumnitz and Reed, 1937, p. 93).

Period during which price remained constant for a year or more as percent of total months in entire period	Number of cities
70-80 percent	1
60-69	4
50-59	1
40-49	3
30-39	2
20-29	1
10-19	1
Total	13

The margins and earnings of the large dealers were maintained from 1925 through 1929 and nearly so through 1932. The rigidity of wages, fixed by union contracts, held up their costs and they were able to keep margins up (Black, 1935, pp. 81-82).

The farm value (return to the farmer) for fluid milk dropped sharply as the depression worsened, but the spread between farm and retail prices declined much more slowly (Scott, 1957, p. 111).

	Farm value	Farm-retail spread
	- - - -Cents per quart- - - -	
1923-28	6.2	7.0
1929	6.5	7.2
1930	6.2	7.3
1931	5.0	7.0
1932	3.9	6.3
1933	3.7	6.1
	Percent drop from 1929	
1933	43	15

The influx of producer-distributors, peddlers, cut-rate stores, and other distributors, combined with the precipitous drop in effective demand, broke retail milk prices almost completely in most milk markets by 1933. Profits of milk distributors dropped sharply, but in most markets were still positive. Only in two markets investigated by the Federal Trade Commission were they negative on the average for one year from 1930 to 1934 (table 10).

Table 10-- Profits of milk distributors as percent of investment in the milk business, 1930-35

Market and company size	1930	1931	1932	1933	1934	1935
Connecticut:						
Larger companies	19.6	20.7	13.0	7.0	--	--
Smaller companies	13.9	14.4	9.3	.9	--	--
All companies	18.8	19.8	12.4	6.1	--	--
Philadelphia:						
Larger companies	21.1	21.0	15.4	9.2	5.8	--
Smaller companies	14.5	14.3	12.7	7.9	9.4	--
All companies	21.0	20.8	15.3	9.1	5.9	--
Baltimore:						
2 companies	25.6	--	--	18.1	14.0	14.9
Cincinnati:						
4 companies	13.1	--	--	*.2	8.5	10.0
St. Louis:						
3 companies	19.4	--	--	3.7	.5	*2.8
Boston:						
2 companies	20.5	--	--	13.3	3.3	11.4

*Loss.
Source: Gaumnitz and Reed, 1937, pp. 185-6, from Federal Trade Commission.

MARKETING AND PRICING BUTTER AND CHEESE

With thousands of small plants producing butter and cheese, there was some resemblance to an atomistic market. Marketing channels often involved a number of firms. Most plants produced a bulk product which was often packaged by an assembler who resold to wholesalers or chainstores. (See Nicholls, 1939; Sprague, 1939.)

The pricing of both butter and cheese was, and still is, by a quotation system based on limited transactions on the exchanges in Chicago and New York for butter and in Green Bay (now in Plymouth), WI, for cheese. Most sales by the plant and at later levels in the channel were made at established differentials from the quotation.

MARKET ORGANIZATION AND POWER

Eight large dairy companies played an important role in market for dairy products for most of this period. Several date back into the 19th century, but the major growth of all occurred after the turn of the century and all but one since the mid-twenties. Much of their growth--like that of other industrial firms throughout the country--occurred during two of the three merger movements in the United States.

The first wave of industrial mergers around the turn of the century passed the dairy industry by. The second merger movement--during the twenties--saw the Borden Company more than double its sales from over $100 million in 1919, primarily by mergers within the dairy industry. The National Dairy Products Corporation was organized in 1923 and immediately began a period of rapid growth, primarily through mergers. By 1930, it had become the largest company in the dairy industry. Beatrice Creamery Co. and Fairmont Creamery Co. were also organized in the early twenties, but grew relatively slowly until after World War II, when they too engaged in numerous mergers.

In 1934, the three largest companies had national market shares (sales--not production) of--

23 percent of fluid milk
19 percent of butter
63 percent of natural cheese
44 percent of canned milk   (U.S. Fed. Trade Comm., 1937)

If all of the dairy operations of large companies--processing of fluid milk and manufactured products, as well as purchases of products--are aggregated, 22 companies handled 39 percent of all milk and cream products which left the farm in 1934 (table 11). The 12 large dairy companies (which includes A&P) accounted for 27 percent and 10 meat packers for 12 percent. The largest four companies handled 26 percent of all dairy products (in terms of milk used in those products) and the next four another 8 percent.

Information on rates of return on capitalizaton for four large dairy companies are available from the mid-twenties through the late thirties. These figures cover all of the operations of the companies, although in this period they were heavily concentrated in

Table 11--Percent of sales of milk and cream to plants and dealers
represented by purchases and handlings of 22 large firms, 1934

Company	Percent of sales of milk and cream to plants and dealers
Dairy companies:	
National Dairy Products Corp.	10.6
The Borden Co.	7.6
Beatrice Creamery Co.	3.4
The Fairmont Creamery Co.	2.8
Carnation Co.	1.4
Pet Milk Co.	1.1
Golden State Co., Ltd.	1.1
The Great Atlantic and Pacific Tea Co.	.7
American Dairies, Inc.	.6
Western Dairies, Ltd.	.5
North American Creameries	.3
Creameries of America, Inc.	.2
Subtotal	30.5
Meat packers:	
Swift and Co.	5.5
Armour and Co.	5.3
Cudahy Packing Co.	1.1
Wilson and Co., Inc.	1.1
Kingan and Co., Inc.	.4
John Morrell and Co., Inc.	.2
Jacob Dodd Packing Co.	.1
Hygrade Food Products Corp.	.1
George A. Hormel and Co.	*
The Rath Packing Co.	*
Subtotal	13.7
Total 22 companies	44.2

*Less than .05 percent.

Source: U.S. Fed. Trade Comm., 1937, p. 227.

dairy operations. Rates of return ranged from 16 to 18 percent in the late twenties, dropped to a low of 5 percent at the bottom of the depression, and began to recover in the late thirties (table 12). Figures for 10 companies for a part of the period average somewhat lower than for the four large companies, but show the same general pattern.

In 1932, while nobody felt as though he had much market power in a demoralized marketplace, what existed was clearly lodged in the hands of the processors or distributors. In butter and cheese, the strongest hand was held by the assemblers and packagers. It was they who applied the brands, controlled quality standards, and made the prices. In short, they were the funnel through which the product had to pass from several thousand plants to several hundred thousand retailers. Land O'Lakes and Challenge were working their way into the butter business but were far from equal to the national dairy companies and meat packers.

Table 12--Rate of return on capitalization of large dairy companies, 1925-1937 1/

Year	Four leading companies	Ten dairy companies
1925	16.9	--
1926	17.1	--
1927	16.1	--
1928	18.0	--
1929	18.4	15.7
1930	16.2	14.2
1931	12.9	12.3
1932	7.1	6.3
1933	4.9	4.4
1934	5.7	5.1
1935	6.9	--
1936	10.0	--
1937	8.2	--

1/ Earnings (interest on bonded indebtedness, additions to net worth, Federal income taxes, and dividends) as percent of capitalization (capital stock, bonded indebtedness, surplus and surplus reserves).

Source: Froker et al, 1939, p. 17, from Federal Trade Commission.

Canned evaporated milk was much more important then than now. The power was in the hands of the national companies with their brands, although A&P and a few other grocery chains had their own brands and others with "bargain brands" were a threat.

The power in fluid milk markets rested largely in the hands of the processors, especially the large firms. With most of the milk home delivered by their employees or those whom they supplied, they had only to deal with unorganized consumers and their fellow milk dealers. Monthly meetings of the local milk dealers association or bottle exchange had helped to bring about in many markets the kind of understanding that kept prices stable for long periods of time. While the conditions of 1932-33 were more chaotic, power had clearly not passed to the cooperatives.

Patent control played a part in the early development of the condensed and evaporated milk industries and the processed cheese industry. However, it no longer plays such a role. The growth of Borden and Pet in the condensed and evaporated milk industries was undoubtedly assisted by patent control, but their effect is now only of historical interest (Hoffman, 1940, p. 125). Kraft's dominant position in the processed cheese industry was achieved very largely through patent control. In the thirties, its patents gave it exclusive control over methods of making processed cheese. While patents are no longer effective, the company still dominates the business through aggressive merchandising and product differentiation.

IN SUMMARY

In the thirties, milk production was largely a small-scale, hand-labor enterprise. Only about three-quarters of the milk was sold either as whole milk or farm-separated cream. Assembly of milk was mostly by truck on organized routes, although many farmers with only a few cows still hauled their own (and perhaps their neighbors') milk to the butter or cheese plant or the receiving station. In large markets, much milk for fluid use moved to the city by rail but, in smaller markets with more compact milksheds, all or most was hauled by truck.

Cooperatives were important in butter production but much less important in cheese. In most major fluid milk markets, cooperatives bargained with large fluid milk processors but many small processors dealt with unorganized producers. Either classified pricing or a base-excess plan was generally used by fluid milk cooperatives but smaller processors generally used flat pricing.

Producer-distributors were the major source of fluid milk products in smaller towns and cities and even in the larger cities in the South, selling about a quarter of the fluid milk nationally. In larger cities, from a half to 80 percent of fluid milk distribution was in the hands of three or four larger dealers.

About three-fourths of all fluid milk was home delivered. During the Depression, substantial numbers of people went into milk distribution in an effort to make a living when jobs were essentially unobtainable, but large dealers continued to dominate. The largest companies remained profitable through the thirties, although much less so than in the twenties.

5
Changes in
Production and Assembly

The entire national economy has changed in the past half century...in many dimensions one can find hardly any trace of 1933.

o Population rose 85 percent.
o Gross national product multiplied six times in real terms.
o Price levels rose nearly five-fold.
o Income per person rose three and one-half times in real
   terms.
o Household size dropped 30 percent.
o The economy became multi-national.
o Large firms became larger and more numerous.
o The economy climbed out of the Great Depression and, after
   World War II, economic growth interrupted by occasional
   recessions became the pattern.

The dairy industry experienced drastic change, along with the
rest of the economy. In Chapters 5-7, the dimensions of change in
the dairy industry will be examined as a prelude to looking at
changes in the public role in the dairy economy.

PRODUCTION

Milk production, which had once been nearly universal on farms
in the United States, became largely a specialized form of commercial
farming. In 1930, over 70 percent of all farms had milk cows, but
less than 14 percent of the farms with milk cows were commercial
dairy farms (table 13). The share of farms with milk cows rose during the depression to more than three-fourths and began dropping more
and more rapidly after World War II. By 1978, only 13.5 percent of
all farms had milk cows and nearly half of those were commercial
dairy farms.
Production solely for home use remained important, in terms of
number of farms, until 1955. The number of farms with one or two
dairy cows--nearly all of whose output was used at home--was remarkably stable at about 2.3 million through the thirties and the first
half of the forties. After World War II, the 1-2 cow operations

Table 13--Number of farms

Year	:All farms:	Farms with: milk cows	Home use: only 1/	Selling dairy products 2/	Dairying as sideline 3/	Commercial dairy farms 4/
			Thousand farms			
1929 :	6,295	4,453	1,570	2,883	2,289	594
1934 :	6,812	5,237	---	---	---	---
1939 :	6,102	4,644	2,013	2,631	2,012	619
1945 :	5,859	4,495	2,022	2,473	1,914	559
1950 :	5,382	3,682	1,675	2,007	1,405	602
1954 :	4,782	2,957	1,482	1,475	926	549
1959 :	3,711	1,842	825	1,017	589	428
1964 :	3,158	1,134	493	641	274	367
1969 :	2,730	568	208	360	99	261
1974 :	2,450	404	157	247	51	196
1978 :	2,479	334	113	221	53	168
1981 :	2,419	324	---	---	---	---

1/ Farms with milk cows but selling no dairy products.
2/ Farms selling milk, cream, or butter, 1929-1959. Farms selling milk or cream, 1964. Farms with total sales of $2,500 or more which sold some dairy products, 1969 and 1974.
3/ Farms which sell dairy products but are not commercial dairy farms.
4/ Definitions vary from one Census to another. 1940 on same basis as 1945, 509,000.

Source: 1930-1978: Census of Agriculture. 1981: Crop Reporting Board.

dropped rapidly until there were only 124 thousand such farms in 1974. About a quarter of all milk was used on farms where produced in the thirties, dropping to 7 percent in the sixties and only 2 percent today (table 14).

Farming of every kind became much more specialized during these years. The number of general farms and other types of specialized farms which had milk cows declined from over 2 million in 1930 to 166 thousand in 1978. Farms with 3-9 cows--largely sideline operations-- declined from 39 to 32 percent of all herds between 1929 and 1944 and to 14 percent by 1974. Farmers with 5-9 cows were 5 percent in 1978.

Herd size increased throughout the period on larger farms (table 15). Farms with 30 or more cows accounted for 9 percent of milk production in 1929, 12 percent in 1939, 16 percent in 1944, 58 percent in 1964, and 81 percent in 1975. The farms with 100 or more cows accounted for only 1 percent of all farms with milk cows in 1964 but produced 13 percent of the milk. By 1975, 3 percent of the farms produced 24 percent of the milk.

Table 14--Milk production on farms and disposition from farms, 1930-1981

| Year | Production | Used on farms where produced | Total | Sold | | Retailed |
| | | | | To plants and dealers | | |
				Whole milk	Cream 1/	
			-Billion pounds-			
1930	100	25	75	34	34	7
1935	101	26	75	36	33	7
1940	109	23	86	47	33	6
1945	120	21	98	69	24	6
1950	117	18	98	74	20	4
1955	123	15	108	91	15	3
1960	123	9	114	104	8	2
1965	124	6	118	113	4	2
1970	117	4	113	110	1	2
1975	115	3	112	110	*	2
1980	129	2	126	125	*	1
1981	133	2	130	129	*	2

*Less than 500 million pounds.
Detail may not add to totals because of rounding.

1/ Milk equivalent of farm-separated cream.

Rural transportation systems improved as rural roads were paved and the selling of farm-churned butter, which had been common on isolated farms with a few cows in the twenties and thirties, nearly disappeared. In 1930, 45 percent of farm sales of milk and cream were of farm-separated cream. The amount, but not the percentage, held constant through the thirties and dropped sharply during World War II as the Federal government encouraged utilization of the skim milk portion of farm milk in nonfat dry milk and rising demand helped to make markets for it. Since then, the sale of farm-separated cream has nearly disappeared.

Milk production per cow dropped 10 percent between 1930 and 1935, as the number of cows increased, grain feeding decreased, and droughts affected pastures. By 1940, it had risen above the 1930 level and, since the end of World War II, the rate has risen almost continually...

	Milk per cow	Annual average change
	Pounds per cow per year	
1930	4,508	+58
1935	4,184	−65
1940	4,622	+88
1945	4,787	+33
1950	5,314	+305
1955	5,842	+106
1960	7,029	+237
1965	8,304	+255
1970	9,751	+289
1975	10,360	+122
1981	12,147	+298

In the thirties, milk production was largely a hand operation (see table 4). The number of man-hours per cow was about the same as before World War I. Power equipment for milking and care of the cows was unavailable on most farms until rural electric firms strung power lines through the farm country. Since World War II, both the dairy operation and field operations have become increasingly mechanized. In the seventies, only 0.4 hour of labor was used to produce a hundredweight of milk, compared to 3.4 hours during the thirties (table 16). The increases in labor productivity in field work were even more dramatic--from 9.5 to 1.5 hours per ton of hay.

Table 15--Size distribution of farms with milk cows

Number of: milk cows:	1929	1939	1949	1959	1969	1974	1981
	- - - - - - - - - - Percent of farms - - - - - - - - - - - -						
1-9	88.0	87.4	82.2	71.3	49.0	44.6	
10-19	9.8	10.0	12.9	14.2	14.8	11.0	58.7
20-29	1.6	1.8	3.2	7.7	12.9	11.7	
30-49	.5	.7	1.3	4.9	14.8	17.7	19.5
50-99	.1	.1	.3	1.5	6.8	11.4	16.1
100 or more	*	*	.1	.4	1.7	3.6	5.7
Total	100.0	100.0	100.0	100.0	100.0	100.0	100.0

* Less than 0.05 percent.

Sources: 1929-74: Census of Agriculture. 1981: Crop Reporting Board.

Table 16--Labor productivity in milk and hay production

Period	Man-hours per cow	Milk per cow (cwt.)	Man-hours per cwt. of milk	Man-hours per ton of hay
1930-34	147	43	3.4	9.5
1935-39	148	44	3.4	9.1
1942-44	142	47	3.1	8.1
1945-49	129	50	2.6	6.2
1950-54	121	54	2.2	4.4
1955-59	109	63	1.7	3.7
1960-64	93	75	1.2	2.8
1965-69	78	88	.9	1.9
1976-80	42	113	.4	1.5

Milking machines are now found on every dairy farm, except for those of Amish farmers in Pennsylvania and elsewhere who refuse to use electric power for religious reasons. Forage harvesting is highly mechanized. Feed handling changed from bagged to bulk feed in the sixties and seventies. In the Northeast, bulk feed increased from 22 percent of formula feed production in 1969 to 66 percent in 1975 (Vosloh, 1971 and 1978).

Bulls have largely disappeared from dairy farms, as artificial insemination became general. Horses too are gone from dairy farms, with tractor power replacing them in field work. This released man-hours for other work, as well as allowing feed and pasture to be used entirely for dairy production.

The result of all these changes is that a one-man farm produces about seven times as much milk as it did during the thirties and three or four times as much as during the fifties.

The seasonality of milk production has been drastically reduced since the mid-fifties (table 17). Production was about half again as large in the peak month of May or June as it was in the low month of November during the thirties. In the late forties, seasonality actually increased a bit because fall production dipped a little lower. Since 1955, seasonality has declined so that the peak is now only 12 percent greater than the low month, with the change occurring both in Spring flush and Fall.

Incentives to reduce seasonality of production have been provided by prices. Reflecting seasonal differences in supply/demand relationships, class prices have been higher in the fall than in the spring throughout this period. In addition, blend prices to Grade A producers show greater variation than class prices because added production in the spring goes almost entirely into lower-valued manufactured products. For many years, a number of Federal orders included seasonal payment plans to further encourage seasonal leveling of production. Quota plans of some cooperatives and State milk

Table 17--Seasonal variation in milk production

Year	:	Production in--		:	Peak month : as percent of low month
	:	Peak month	: Low month	:	
	:		:	:	
	:	Percent of monthly average			
	:				
1930	:	127	83		153
1935	:	127	85		149
1940	:	127	81		157
1945	:	129	78		165
1950	:	127	80		159
1955	:	125	83		154
1960	:	120	87		138
1970	:	113	91		124
1981	:	107	95		112
	:				

Peak month was June in 1930-47 and May in 1948-81.
Low month was November throughout.

control agencies (as well as the Federal order plan in the Puget Sound market) almost universally use the low production months as the base-building period.

Milk production has shifted toward the areas of specialized production--the Northeast, the Lake States, and the Pacific Coast (table 18). These three regions accounted for 47 percent of the national output in 1929 and 1939, but the share increased to 63 percent by 1981. The regions with large declines were the Corn Belt and the Northern Plains. Production in the Corn Belt held at 21 percent of the national total until 1949, with the biggest decline since 1964. Production has been declining in the Northern Plains since 1929.

The reasons for the shifts between regions are straightforward production economics. The increasing specialization of dairy farming and all other forms of the farm enterprise have concentrated milk production where it has a comparative advantage--for fluid use, production is generally located near the point of consumption, because of the high cost of transport. Production for manufactured products is increasingly concentrated in the three regions where the comparative advantage of milk production is greatest--the Northeast, the Lake States, and the Pacific Coast. In the seventies, comparative advantage in crop production accentuated the relative decline of milk production in the Corn Belt. The Lake States and the Northeast, where comparative advantage remains with milk, have maintained their share, with the big increase on the Pacific Coast with its large dairy farms.

Table 18--Milk production by region

Region	1929	1939	1949	1964	1981
	Percent of United States				
Northeast	17	17	18	20	20
Lake States	23	23	24	29	29
Corn Belt	21	21	21	18	12
Northern Plains	10	9	7	6	4
Appalachian	7	7	8	7	7
Southeast	3	3	3	3	3
Delta States	3	3	3	2	2
Southern Plains	6	6	5	3	4
Mountain	4	4	4	4	5
Pacific 1/	7	7	8	9	14
United States	100	100	100	100	100

Detail may not add to total due to rounding. Regions as in figure 1.

1/ Includes Hawaii and Alaska.

Certainly, too much was produced in 1981 and 1982 in high-cost regions--which might lead an observer to wonder if the wrong price signals were being sent to milk producers. But too much milk was produced nearly everywhere, in low-cost and high-cost areas, which would seem to indicate that the overall price level was too high and its effect was to encourage milk production beyond needs at going prices in all areas.

MARKETING BULK MILK

This section is mostly about the marketing of bulk fluid grade milk from farm to processor. Many of the technological changes in the marketing and movement of fluid grade milk also took place for manufacturing grade milk, although not always at the same pace. But, since manufacturing plants were located much closer to the farm, the process was much simpler.

As we have seen earlier, only about a third of the milk produced in the early thirties was sold to plants and dealers as whole milk (table 2). Most of the milk of fluid grade--80-85 percent--was used in fluid milk products (table 19). Increases in the demand for fluid milk products as the country recovered from the depression and then boomed during World War II brought forth increases in the production of fluid grade milk. Increasing demand for manufactured products brought a massive switch from farm-separated cream to whole milk and the sale of manufacturing grade whole milk doubled from 1940 to 1955, where it leveled off for ten years, and then dropped.

About three-fourths of all whole milk sold to plants and dealers

Figure 1
Milk Production By Region, 1929 and 1981

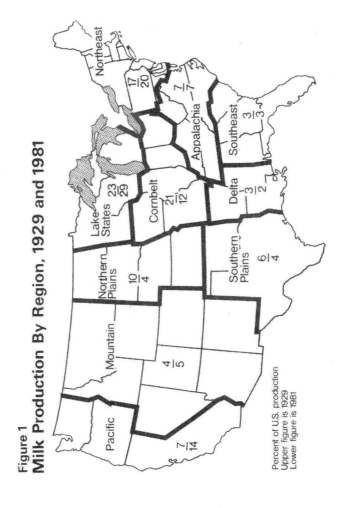

Percent of U.S. production
Upper figure is 1929
Lower figure is 1981

Table 19--Utilization of milk sold to plants and dealers 1/

Year	Total	Manufacturing grade milk used in manufactured products	Fluid grade milk used in--	
			Manufactured products	Fluid milk products
		- - - - - - - - - - Billion pounds - - - - - - - - - - - -		
1930	34	9	4	22
1932	34	8	4	21
1935	36	12	3	20
1940	47	17	6	24
1945	69	30	5	35
1950	74	29	8	38
1955	91	34	12	45
1960	104	34	18	51
1965	113	35	24	54
1970	110	29	31	50
1975	110	22	38	50
1980	125	20	54	50
1981	129	19	61	49

1/ Sum of the parts may not equal totals because of rounding. Before 1950, the estimates of milk by grade are rough approximations by the author.

was fluid grade in the early thirties. The share declined through the late thirties and early forties, reaching about 57 percent in 1945. Since then, it has increased in nearly every year reaching 63 percent in 1955, 69 percent in 1965, and 85 percent in 1981.

Manufacturing grade milk production in the postwar period was heavily concentrated in the Upper Midwest, with the largest quantities in Minnesota and Wisconsin but accounting for most of the milk in the Northern Plains states. There were also significant quantities in the mid-south.

Higher prices for fluid grade milk than for manufacturing milk provided much of the incentive for farmers to convert, although other factors played a part (table 20). These are the differences between state-wide averages for the two grades of milk, not the differences facing a given farmer. In the sixties and early seventies, Graf and Jacobson estimated that a difference of 15 cents per hundredweight between the Grade A blend and the manufacturing grade milk price was sufficient to encourage conversion to Grade A. The increase in Class 1 differentials in 1966 extended the area where such order blend price differentials existed about 100 miles further into the Chicago milkshed (Graf and Jacobson, 1973, p. 60). The other factors, including bulk tank assembly, stricter standards for manufacturing grade milk, and plant efficiencies, are all related to the entire assembly and country plant system, to be discussed shortly.

Table 20--Average prices received by farmers for milk sold to
plants and dealers: Difference between fluid grade and
manufacturing grade, Minnesota and Wisconsin

Period	:	Minnesota	:	Wisconsin
	:	Cents per cwt. for 3.5 percent milk		
1960-64	:	33		43
1965-69	:	40		43
1970-74	:	47		42
1975-79	:	56		46
1980-81	:	63		54

Assembly and movement of milk to market were significantly
changed by technological developments during the past 50 years. By
1932, trucks had largely taken over the assembly of milk from farms
to nearby receiving stations or country plants, but long-distance
movement of milk was still dominated by rail transportation. Highway
networks were improved, so that faster movement became possible.
Rural roads were increasingly paved and primary highways improved,
especially with the development of the Interstate highway system
after 1956 (table 21). Between 1946 and 1950, long distance movement
of bulk milk shifted from rail to truck (U.S. Agric. Mktg. Serv.,
1955, p. 88).

Hundreds of receiving stations and country plants dotted all but
the most compact milksheds in the early thirties. Early studies had
clearly demonstrated substantial economies of scale in the operation
of receiving stations and country plants (Bressler, 1942). As road
transportation improved and farms selling whole milk became larger,
milk collection routes were extended and receiving stations consolidated. The remaining receiving stations became larger. Country
plants which operated both as receiving stations, taking in cans of
milk from the farm and consolidating them into tankloads of milk for
market, and as manufacturing plants for the milk not needed in the
city for fluid products, took on an increasing share of the volume.

In smaller markets, receiving stations disappeared and most milk
was hauled directly from the farm to city plants.

The technological development with the most dramatic impact on
bulk milk marketing was the farm bulk tank. It changed the entire
procedure of handling milk from the farm to the processing plant.

Milk had been put into 40-quart cans on the farm for at least a
half century. These were then cooled, either with ice and water or,
as electricity came to the farms, with mechanical refrigeration.
The cans were hauled to the receiving station or plant and then
dumped. With the introduction of mechanically refrigerated bulk
tanks on the farm, milk was cooled promptly, pumped into a pickup
tank truck, hauled to the plant, and pumped out.

Table 21--Rural roads and truck transportation

Year	:	Mileage of paved rural highways	:	Farms on hard-sur- faced road	:	Average speed of trucks	:	Miles of rural travel by trucks
	:	Percent paved		Percent		Miles per hour		Bill. vehicle miles
	:							
1925	:	10		7		--		--
1930	:	16		9		--		--
1933	:	23		--		--		--
1935	:	28		--		--		--
1936	:	--		--		--		22
1940	:	37		19		--		30
1945	:	49		--		39.8		27
	:							
1950	:	56		--		43.0		57
1955	:	64		--		45.6		71
1960	:	70		--		48.2		82
1965	:	72		--		51.8		103
1970	:	76		--		54.7		134
1975	:	78		--		54.8		157
	:							

Sources: U.S. Bu. of Census, 1978, pp. 629 and 634; U.S. Bu. of Census, 1975b, pp. 710 and 718.

Bulk tank handling on the farm was first introduced in California in the late thirties on a few farms with several hundred cows apiece where an entire truck tankload of milk was produced at one milking (Beal and Bakken, 1956, p. 175). Later, it spread to similar large farms in Florida, but use on smaller farms awaited development of refrigerated tanks that cooled milk quickly at reasonable cost on farms producing less than 300 gallons per day (Stocker, 1954). An experimental route was established in Connecticut in 1948 and spread to other States in 1949-51 (Beal and Bakken, 1956, p. 176). By 1953, 6,150 farms had bulk tanks, about three-fourths on farms producing fluid grade milk and one-fourth on manufacturing milk farms (Cowden, 1956).

The spread was rapid as both handlers and producers saw the economies to be achieved. Processors and cooperatives helped to arrange financing for farm tanks in order to speed up the conversion process. Premiums were paid to farmers using bulk tanks. Growth was rapid... (Stocker, 1954; Cowden, 1956; Herrmann and Agnew, 1959; Agnew, 1957; Dairy and Food Ind. Supply Assoc., 1966 and 1967).

March 1955... 18,000 farms with bulk tanks
Dec. 1956.... 57,000 farms with bulk tanks
1958......... 60,000 farms with bulk tanks
1963........ 200,000 farms with bulk tanks
1965........ 218,000 farms with bulk tanks

Essentially all fluid milk farms are now equipped with bulk tanks. Manufacturing milk areas are also converting at a fairly rapid rate. By 1975, 58 percent of Minnesota Grade B farmers with 75 percent of the output had bulk tanks (U.S. Stat. Rpt. Serv., 1977, pp. 14-15).

The shift to bulk milk handling increased the investments of dairy farmers, lowered the costs of moving milk from farm to plant, made many country milk handling facilities obsolete, and made the entire system much more flexible. No longer tied to fixed country locations, milk could be moved where needed much more easily. This opened up opportunities for changes in organizational forms.

Cooperative Role

All of these technological and economic developments, bulk tank handling perhaps more than the others, changed the economic opportunities and constraints facing cooperatives and handlers in the market for bulk fluid milk. Cooperatives had grown from about 45 percent of the bulk fluid milk market in 1927 (Metzger, 1930) to about 54 percent in 1936 (Hirsch, 1951). By 1957, 735 cooperatives sold 62 percent of the total (Gessner, 1959). Very nearly the same number of cooperatives handled more milk in 1964, but that was only 57 percent of the larger total. In 1973, mergers had reduced the number of cooperatives to 458, but they handled 63 percent of the total—almost the same as in 1957 (Tucker et al, 1977).

In the twenties and thirties, cooperatives in the bulk fluid milk market were largely confined to bargaining. Most of the physical handling of the milk was done by others, either the processors or, in the case of hauling, independent truckers under contract with them. Some country plants and receiving stations were operated by cooperatives, but the great majority were owned by the processors. Typically, the cooperative bargained over prices (where they were not set by Federal or State regulation) and other terms of trade. Each processor received all the milk from farms on specified assembly routes which produced approximately the amount of milk needed by that processor. The processor was responsible for all the milk and for the disposition of that part of his receipts which was not used for fluid milk products.

In the larger markets, where most milk (except that produced in nearby areas) went to receiving stations or country plants, smaller processors without sufficient volume to operate a receiving station purchased from independent receiving stations, country plants, or a larger processor.

From the late thirties to the middle fifties, there was some shift of country plant operation from processors to cooperatives. Then, as the shift to bulk handling of milk accelerated, a major shift in the roles of cooperatives and processors began. Cooperatives took over much more milk hauling. With the increased mobility of milk, it could move directly from the farm—perhaps with a pause at a pumpover station for transfer to a larger tank truck—anywhere in the market or even to a nearby market. It became evident that better management of the movement of milk could bring significant reductions in costs and that economies of scale were substantial. Research in the sixties provided measures of the potential cost savings (Lasley, 1964 and 1966). Total economies in such a centralized supply-coordination and surplus-disposal operation, compared with a system in which each handler managed his own supply and surplus disposal, were about 20-22 cents per hundredweight at that time (Manchester, 1971, p. 6). With inflation, they have much more than doubled since then.

The size of fluid milk plants grew during the sixties with small plants closing, middle-sized plants adding volume, and the construction of some very large plants (see Chapter 6). The very large plants--some built by supermarket chains and others by proprietary firms (often consolidating a number of plants)--were a response to economies of size which reflected the new technologies and to the increased mobility of milk.

These large plants were specialized fluid milk plants--they had little or no capacity to handle daily or seasonal surpluses. In other words, they were planned to utilize a procurement system operated by a large full-service cooperative. In some cases, local cooperatives were not large enough to provide the volume needed by such a plant, providing incentive for mergers among cooperatives to achieve such a size.

Transport costs have risen less than milk prices in the past twenty years. The cost of moving bulk milk on long hauls declined from about 3 percent of the U.S. average Class 1 price to about 2 percent. Thus, movement of milk within and between markets became less costly in a relative sense.

Since then, many processors have entered into a full supply arrangement with a cooperative because of high costs of procuring and coordinating a fluctuating supply to meet a variable demand and the possibility of eliminating some uncertainty in this area. Under such a full supply arrangement, the cooperative undertakes to supply the exact needs of the processor for milk for fluid use and perhaps for ice cream and cottage cheese, and also to dispose of the surplus for other uses. Milk supply varies from day to day, depending on the vagaries of production by individual cows, weather, road conditions, and other uncontrollable factors. Demand likewise varies from day to day, partly on the basis of the day of the week, since more and more milk is being sold through supermarkets with a concentration of sales on weekends. Also, there is a strong element of random variation in both supply and demand from day to day. The larger the volume under the control of one agency, the more the random variations tend to offset one another, both within supply and demand and between the two.

The day-to-day variation in demand for milk by fluid milk bottling plants is indicated by the following tabulation of receipts of milk each day at individual plants relative to those on the highest-volume day in each of the months of May and October (Lasley and Sleight, 1979, p. 14):

Receipts as percent of high day	Percent of days
81-100	25
61-80	26
41-60	20
21-40	11
1-20	4
0	14

Day-to-day variation in supply is equally wide. Receipts of seven plants in Pittsburgh in March, May, September and December 1963, when each plant largely handled its own supply from a given set of producers, illustrate (Lasley, 1966, p. 16):

Percent of average daily receipts for that month	Percent of days
74 or less	8
75-89	18
90-114	58
115-129	13
130 or more	3

A full supply arrangement does not adjust supply or eliminate fluctuations, but it does reduce their impact on the processor by giving him a relatively simple, routine means of adjusting supply to demand with minimum effort and expense. Furthermore, a single agency is better able to make necessary adjustments and reduce the burden of uncertainty.

These functions, formerly performed by processors, are largely performed by full-service cooperatives today. Significant costs formerly borne by the processor are now borne by the cooperative and total costs are reduced. This has increased efficiency in milk marketing and, at the same time, cooperatives have grown and acquired increased market power.

The cooperative share of all milk marketed increased from 59 percent in 1957 to 76 percent in 1973 (table 22). The increase was evenly divided between bulk whole milk sales and manufacture of dairy products.

Intermarket Shipments

Before 1941, large cities in the United States were usually able to obtain their full fluid milk requirements from local farms at all seasons. Demand for milk first outstripped local supplies in the South just before U.S. entry into World War II, when large army camps were established in the region. Increases in milk consumption during and after the war created additional shortages in the South, the Northeast, and elsewhere. Even in markets where total annual production equaled or exceeded requirements, many found themselves short of milk each year during the season of low production (U.S. Agric. Mktg. Serv., 1955, p. 6). In part, these shortages were due to a policy of the War Food Administration of allocating feed supplies to the

Table 22--Cooperatives, 1957, 1964, and 1973

	:	1957	:	1964	:	1973
	:		:		:	

	Million pounds		
Milk handled by all cooperatives 1/	58,038	76,743	2/ 83,227
Raw whole milk sold by all cooperatives 3/	36,213	43,443	52,180

	Pct. of all milk sold to plants and dealers		
Milk handled	59.1	67.2	75.8
Raw whole milk sold	36.8	38.0	47.5

	Pct. of milk handled by cooperatives that is sold as raw whole milk		
	62.4	56.6	62.7

1/ Milk received by cooperatives and that sold by cooperatives and delivered directly to other handlers. Includes small amounts purchased from noncooperatives (0.8 percent in 1973), but excludes substantial purchases from other cooperatives.
2/ Receipts from producers were 83.6 percent Grade A.
3/ Excludes sales to other cooperatives.

Sources: Gessner, 1959; Tucker et al, 1977.

Midwest, thus holding down production in the South and Northeast.

Intermarket shipments increased sharply from 1940 to 1946 and stayed at fairly high levels in 1947 and 1948. Then they dropped as production increased in the South and Northeast. In the Midwest, they maintained a constant share (Lee, 1950).

"Outside" and "Emergency" Milk as
percent of Class 1 sales

Year	Boston, New York, Philadelphia	Omaha, Kansas City, Cincinnati, St. Louis
1940	11	11
1941	11	10
1942	13	--
1943	18	11
1944	16	10
1945	17	10
1946	17	11
1947	14	11
1948	14	--
1949	12	--

Size of Bulk Milk Markets

Until the sixties, bulk milk markets were fairly distinct and relatively well-defined. Producers delivered milk to a cooperative or a proprietary handler who did business in a single market in most cases. In the sixties and seventies, the changes in the functions performed by cooperatives and proprietary handlers that have been discussed in this chapter and the broadening of the distribution areas to be discussed in the next chapter brought about a quantum growth in the size of bulk milk markets. To a considerable extent, bulk milk markets became so highly interrelated that they could only be described in regional terms.

IN SUMMARY

Since the thirties, milk production has become a commercial enterprise, although still on family farms in most parts of the country. Nearly all milk is sold as whole milk, with use on farms and sales of farm-separated cream only a small portion of the total. Farms are much larger and mechanized so that one man can handle three or four times as many cows with the average cow producing several times as much milk.

Milk production has moved increasingly to the areas where it has comparative advantage--the Lake States, the Northeast, and the West Coast. At the same time, more and more milk meets the standards for use in fluid products with the remaining manufacturing grade milk concentrated in the upper Midwest.

The marketing of raw milk changed drastically with introduction of bulk tanks on farms so milk could be picked up by tank trucks and hauled directly to the city from substantial distances. Conversion to bulk handling of milk for fluid use was completed in the sixties and many manufacturing milk plants converted in the seventies. The change to bulk handling and improvement of the highway network made it feasible for cooperatives to increasingly assume the function of supplying fluid milk markets, routing the movement of the milk, and balancing supply and demand. This includes manufacturing the milk not needed into products such as butter, powder, and cheese. There are substantial economies of size in the function, so cooperatives are able to carry it out at lower cost than if each processor did it for himself.

6
Changes in Processing

FLUID MILK PRODUCTS

The processing of fluid milk products shifted from one-fourth by producer-distributors and three-fourths by commercial processors in the early thirties to one-fifth by producer-distributors in 1940 as the country made some recovery from the Depression and producers concentrated on production rather than on distribution (table 23). During and after the war, the share of producer-distributors dropped sharply:

1940	21 percent by producer-distributors
1945	14 percent by producer-distributors
1950	9 percent by producer-distributors
1955	5 percent by producer-distributors
1960	4 percent by producer-distributors
1965-81	3 percent by producer-distributors

As the economies of size curve for milk processing shifted following World War II, small producer-distributors were increasingly at a cost disadvantage. Today, a substantial proportion of the business of producer-distributors is by large operators who have one or more milk stores. This is in decided contrast to traditional producer-distributors with almost all of their sales home-delivered or to small stores and restaurants in the smaller towns.

Rough estimates of the numbers of producer-distributors are as follows...

	Number	Average annual sales (1,000 pounds milk equivalent)
1936	35,600	190
1939	28,650	220
1940	29,500	210
1948	11,700	390
1964	1,680	1,100
1977	700	2,500

Table 23--Sales of fluid milk products by producer-
distributors and commercial processors.

Year	Producer-distributors	Commercial processors	Total
	- - - - - - - - - Million product pounds - - - - - - - - -		
1930	5,356	16,129	21,485
1932	5,595	16,130	21,725
1935	5,684	16,059	21,743
1940	5,189	19,901	25,090
1944	5,728	32,040	37,768
1945	5,315	34,099	39,414
1950	3,578	35,611	39,189
1955	2,428	42,284	44,712
1960	1,930	47,097	49,027
1965	1,758	51,339	53,097
1970	1,700	52,612	54,312
1975	1,547	54,442	55,989
1981	1,488	53,956	55,444

Sources: Manchester, 1973; O'Connell, 1954; Vial, 1977.

Present-day producer-distributors are more likely to have a herd of several hundred cows--some have several thousand--and specialized outlets such as several dairy stores, in addition to a modern milk processing plant. At the other extreme, there are still a number of raw milk distributors who fill the customer's container, since the health rules prohibit their packaging raw milk.

Commercial Processors

Since the development of city milk distribution began in the last century, a major influence on the numbers and size of processors has been a shift in the cost curve--the relative costs of small firms as compared to larger ones. In the early days, very little happened to fluid milk between the farmer and consumer. Equipment used was simple--a horse and wagon, milk cans, and a dipper were about all that was required. Thus, the costs of the small distributor were not greatly different from those of a large one. The introduction of the glass bottle before the turn of the century was one of the earliest developments causing some shift in the shape of the cost curve. Even simple bottle-filling equipment was expensive when used for a few quarts of milk a day and, as a result, many small distributors went out of business or shifted to buying bottled milk from a larger firm which could afford the equipment.

In the first three decades of the twentieth century, many cities adopted ordinances that required the pasteurization of milk. These requirements increased the costs of small processors compared with those of large ones, and many were no longer able to compete. In the twenties and thirties, the introduction of classified pricing plans forced many small handlers to pay the same price as their larger competitors. Many found it uneconomical to do so and they too went out of business. In the late thirties and forties, the introduction of the paper carton acted to raise the cost levels of these smaller distributors.

High-temperature, short-time pasteurization (HTST) replaced vat pasteurization starting in the thirties, both in fluid milk processing and in other dairy plants where pasteurization was a part of the process. Since World War II, automated systems have replaced the older, manually operated systems. These included cleaned-in-place (CIP) flow lines in the early fifties, automatic controls for CIP systems in 1955, spray cleaning of storage tanks, processing vats and tank trucks, and the air-operated valve for CIP systems. By 1960, completely automated systems were possible and in place in many large plants with better quality control and greatly reduced labor requirements (National Comm. on Food Mktg., 1966, p. 183). They tilted the cost curve even further toward large plants.

No national counts of commercial fluid milk processing plants were made until the author's efforts for the National Commission on Food Marketing (National Comm. on Food Mktg., 1966; Manchester, 1968a). The census did not define fluid milk processing as "manufacturing" until 1954. Before then, fluid milk distributors were counted by the Census of Business, but these figures included both processors and others who bought from them and then distributed fluid milk products to homes, stores, restaurants, and institutions (table 24). In the present study, the author's original series which began in 1948 has been extended back to 1934, making use of data from a limited group of states. Thus, the figures before 1948 are indicative rather than precise.

The number of commercial processors rose fairly rapidly during the thirties and the average plant size also increased (table 25). During World War II, the number of plants began to decline nationally and the average size rose 14 percent per year. During most of the forties, the number of plants was actually increasing in the South and West, as the fluid milk business grew more commercialized in those areas, but declining numbers in the East and Midwest offset the increases elsewhere.

After 1948, the long-term trend to fewer commercial fluid milk bottling plants accelerated. The total number of fluid milk processing plants in the United States declined from 8,527 in 1948 to 5,328 in 1960, an annual average decrease of 3.9 percent. In the first half of the sixties it picked up to 6.8 percent per year and, between 1965 and 1970, to an annual average rate of 9.9 percent. In the seventies, the rate averaged 6.8 percent per year.

Average plant size increased throughout the period, reflecting declining plant numbers and increasing volume. The rate of increase peaked in the late sixties.

Table 24--Milk dealers, 1929-48 1/

	:	1929	:	1939	:	1948
	:			Number		
Establishments:	:					
Retail milk dealers	:	3,990		9,452		4,984
Wholesale milk dealers	:	542		1,647		2,657
Total	:	4,532		11,099		7,641
	:					
	:			Million dollars		
Sales of fluid milk and cream:	:					
Retail milk dealers	:	516		540		1,375
Wholesale milk dealers	:	134		280		1,415
Total	:	650		820		2,790
	:					
	:			Million pounds		
Sales of fluid milk and cream:	:					
Retail milk dealers	:	7,980		9,740		13,840
Wholesale milk dealers	:	2,420		6,080		17,480
Total	:	10,400		15,820		31,320
	:					

1/ Includes both processors and dealers who buy bottled milk from processors and distribute. Retail milk dealers have primarily home delivery business. Wholesale milk dealers sell primarily to stores, restaurants, and institutions. Adapted from the Census of Business.

Table 25--Number and average size of fluid milk bottling plants operated by commercial processors

Year	:	Number of plants	:	Average volume processed	:	Annual average rate of change in--	
	:		:		:	Plants	: Volume
	:		:	Mil. product pounds		Percent	Percent
1934	:	9,400		1.9		---	---
1935	:	9,520		2.0		+1.0	+5.0
1936	:	9,730		2.0		+2.2	+3.9
1940	:	9,950		2.1		+0.5	+0.7
1945	:	8,570		4.0		-2.9	+13.6
1948	:	8,527		3.9		-0.1	-0.4
1950	:	8,195		4.3		-2.0	+5.7
1955	:	6,726		6.3		-3.9	+7.7
1960	:	5,328		8.9		-4.6	+7.2
1965	:	3,743		13.6		-6.8	+8.9
1970	:	2,216		23.7		-9.9	+11.8
1975	:	1,408		38.7		-8.6	+8.5
1981	:	1,032		51.8		-5.0	+5.0

Most of the plants that closed were small. Between 1950 and 1964, the proportion of plants processing less than 5 million quarts daily declined from 93 percent of the total to 64 percent (Manchester and Lasley, 1972, p. 23). Between 1965 and 1970 in a group of Federal order and State markets for which comparable data were available, the total number of plants declined 36 percent. The number of plants decreased in all size groups below 4 million pounds per month and increased in all larger size groups (Manchester, 1974a, p. 6).

Type of Firm. In most of the postwar period, there were seven large national companies in fluid milk processing--

> Borden, Inc.
> Kraft, Inc. (formerly National Dairy Products Corp.)
> Beatrice Foods Company
> Foremost-McKesson, Inc. (formerly Foremost Foods)
> Fairmont Foods Company
> Carnation Company
> Pet, Inc.

The first three operated nationally in the thirties. The other four grew from regional to national status by 1948, primarily through acquisitions of smaller companies. In the seventies, Southland Corporation and Dean Foods Company also grew from regional to national. The seven national firms operated 264 fluid milk plants in 1964 and 196 in 1970. The eight companies had 159 plants in 1980 (tables 26 and 27).

Three of the regional firms of 1934 and 1940 became national in scope by 1948--Carnation, Fairmont, and Pet. Arden Farms and H. P. Hood and Sons have been regional firms since 1934. Arden acquired a controlling interest in Mayfair Markets, a West Coast supermarket chain, in the sixties, later merged the two corporations, and largely left the dairy business in the seventies. Hood was acquired by cooperatives in 1980. Fairmont has become principally a convenience store firm.

Creameries of America, Inc., of Los Angeles, was a regional fluid milk processor in 1934 and continued in that status until it was acquired by Beatrice in 1953.

Golden State Co., Ltd., San Francisco, became regional in scope in the late thirties. It was acquired by Foremost in 1955.

Dean Foods and Hawthorn-Mellody Farms Dairy grew by acquisition into regional status in 1953 and Bowman Dairy Company in 1957. Bowman was acquired by Dean in 1966 and some plants were later divested under FTC order.

Consolidated Foods of Chicago became regional in scope in the early sixties when it acquired a Minnesota fluid milk plant in addition to its Lawson Milk Division. It disposed of the Minnesota plant in 1969 under an divestiture order from the Federal Trade Commission and lost its regional classification. Cumberland Farms, a New England operator of dairy stores, grew into regional status when it opened a fluid milk plant in New Jersey in the early sixties and later one in Florida.

Table 26--Size distribution of fluid milk plants, by firm type, United States, December 1980

December sales volume of fluid milk products, thousand pounds	National firms	Regional firms	Local multi-unit	Local single unit	Regional coopera-tives	Local multi-unit coopera-tives	Local single unit coopera-tives	Integrated supermarkets	Total
				Number					
Less than 100	--	--	3	94	1	--	1	--	99
100-249	--	--	--	106	--	--	2	--	108
250-499	1	--	3	90	--	--	1	1	96
500-999	1	--	3	112	--	2	4	--	122
1,000-1,999	14	--	7	84	3	4	7	1	120
2,000-2,999	17	--	6	44	8	10	2	4	91
3,000-3,999	18	1	3	26	3	--	2	5	58
4,000-4,999	24	1	2	28	3	3	3	3	67
5,000-7,499	25	1	8	36	11	6	3	14	104
7,500-9,999	26	3	3	21	10	3	--	6	72
10,000-14,999	18	5	5	15	6	5	2	10	66
15,000-19,999	5	1	1	7	1	2	--	7	24
20,000-24,999	8	1	3	--	1	2	--	8	23
25,000-29,999	2	--	1	1	--	--	--	4	8
30,000 and over	--	1	1	3	1	--	--	2	8
Total	159	14	49	667	48	37	27	65	1,066

Source: Lough, 1981.

-- = No plants.

Table 27--Number of fluid milk bottling companies and plants,
United States, 1934-1980

Type of firm	1934	1948	1950	1957	1964	1970	1980
	Companies						
National firms	3	7	7	7	7	7	8
Regional firms	6	--	4	5	8	7	4
Local firms:							
Multi-unit	--	--	--	120	99	44	19
Single unit	--	--	--	4,760	3,234	1,609	667
Cooperatives:							
Multi-unit	} 163	--	--	455 {	35	23	18
Single unit					152	83	27
Intergrated super-markets: 1/							
Sole outlet 2/	2	4	6	12	21	25	19
Others 3/	0	0	0	0	2	4	15
Total	8,930	7,750	7,430	5,359	3,558	1,802	777
	Plants						
National firms	--	--	--	--	264	196	159
Regional firms	--	--	--	--	71	57	14
Local firms:							
Multi-unit	--	--	--	--	229	118	49
Single unit	--	--	--	4,760	3,234	1,609	667
Cooperatives:							
Multi-unit	--	--	--	137	115	102	85
Single unit	--	--	--	383	152	83	27
Integrated super-markets:							
Sole outlet 2/	3	10	12	21	36	47	50
Others 3/	0	0	0	0	2	4	15
Total	9,600	8,527	8,195	6,187	4,103	2,216	1,066

-- Not available
1/ Firms whose primary business is operating supermarkets.
2/ Most milk sales through own stores.
3/ Substantial sales through outlets other than own stores. Excludes Arden-Mayfair which is classified as a regional firm through 1970 and local in 1979.

Farmbest (Home Town Foods) emerged as a regional firm in 1966, when it was created to take over the Southeastern Division of Foremost Foods, divested by FTC order. It was acquired by Dairymen, Inc., a regional cooperative, in 1978.

Southland Corp. was originally an ice manufacturer and distributor in Dallas. It entered fluid milk processing in Dallas and the convenience food store business with 7-11 Stores during the thirties. In the early sixties, it began to expand its fluid milk processing business by acquisitions in Texas and Oklahoma, later expanding into other regions.

The many changes in classification resulted in 6 regional firms in 1934, 7 in 1940, 4 in 1950, 5 in 1957, 8 in 1964, 7 in 1970 and 4 in 1980. The 8 firms of 1964 operated 71 fluid milk processing plants and the 7 companies of 1970 operated 57. In 1980, the remaining 4 companies had 14 plants.

Since World War II, multi-unit firms have been consolidating plants. As transportation technology improved, the distribution area from a given plant could expand and it became uneconomical for a large company to operate as many plants as it previously had. Between 1964 and 1971, the number of plants operated by the seven national dairy firms declined from 264 to 188, a decrease of 76. For 56 of these plants, we were able to determine the circumstances surrounding the plant closing. Of this group, 55 percent were cases of plant consolidation. In 29 percent of the cases, the firm withdrew from the market entirely. Another 12 percent were due to FTC-ordered divestitures. In 4 percent of the cases, the company closed its plant and had its product custom-packaged by another processor.

National Market Shares of Large Companies. The three national companies of 1934 accounted for 32 percent of the sales of commercial processors (table 28). With the addition of four other companies to the ranks of national companies and many acquisitions by the original three firms, their total sales increased sharply by 1950, but their share dropped to 22 percent. The acquisitions of the fifties raised it to 29 percent in 1957. The merger policy of the Federal Trade Commission substantially slowed growth of the national firms in the fluid milk business in the late fifties and early sixties and, by 1970, it had been reversed by divestitures under FTC order and withdrawal from unprofitable markets, as the large companies diversified into other lines than fluid milk. With two companies added to the category, the national company share rose to 25 percent in 1980—omitting the two added companies, the share declined.

For five large dairy firms, sales of nondairy products increased from 12 percent of total sales in 1940 to 19 percent in 1950 and 29 percent in 1960 (Bartlett, 1961, p. 1). In the sixties and seventies, diversification increased at a rapid rate and now none of the national dairy companies do as much as half of their business in dairy products.

Table 28--Sales of fluid milk products, by type of firm
1934-1980

Type of firm	1934	1950	1957	1964	1970	1980
	- - - - - - - -Million pounds- - - - - - - - -					
National firms	4,744	7,715	12,927	13,717	12,251	13,579
Regional firms	847	1,487	2,505	2,580	4,038	2,173
Local firms 1/	8,568	23,909	26,304	27,603	25,485	21,021
Cooperatives 2/	700	2,500	3,529	4,896	6,020	8,039
Integrated super-markets: 3/						
Sole outlet 4/	--	--	--	1,471	4,287	7,713
Other 5/	0	0	0	214	438	1,792
Total	14,859	35,611	45,265	50,481	52,519	54,318
	- - - - - - - - - - Percent- - - - - - - - - - -					
National firms	31.9	21.7	28.6	27.2	23.3	25.0
Regional firms	5.7	4.2	5.5	5.1	7.7	4.0
Local firms	57.7	67.1	58.1	54.7	48.5	38.7
Cooperatives	4.7	7.0	7.8	9.7	11.5	14.8
Integrated supermarkets:						
Sole outlet	--	--	--	2.9	8.2	14.2
Other	0	0	0	.4	.8	3.3
Total	100.0	100.0	100.0	100.0	100.0	100.0

1/ Includes integrated supermarket firms, 1934-1957.
2/ Approximations, 1934 and 1950.
3/ Firms whose primary business is operating supermarkets.
4/ Most milk sales through own stores. Included with local firms, 1934-1957.
5/ Substantial sales through outlets other than own stores. Excludes Arden-Mayfair which is classified as a regional firm through 1970 and local in 1980.

Sources: Manchester, 1974a; Lough, 1981.

In 1964, the seven national dairy firms plus Hawthorn-Mellody had the following distribution of sales-- (National Comm. on Food Mktg., 1966, p. 69)

Domestic dairy sales:
Fluid milk products	37	percent
Frozen desserts	12	percent
Manufactured dairy products	18	percent
All dairy products	67	percent
Nondairy and nondomestic sales	33	percent
Total	100	

The five most diversified of the eight companies--National, Borden, Beatrice, Carnation, and Pet--had 58 percent of their sales in domestic dairy products in 1964, compared to 81 percent in 1954. The other three companies were less diversified with 82 percent of sales in domestic dairy lines in 1964, down a bit from 86 percent in 1954 (National Comm. on Food Mktg., 1966, p. 105). By 1975, mergers of Foremost with McKesson-Robbins and of Hawthorn-Mellody into National Industries made the group more diversified with only 23 percent of their sales in dairy products. Dairy product sales of the other five firms were 39 percent of total sales, with the average for all eight firms at 36 percent (table 29).

Table 29--Dairy product sales of eight large companies, 1954-75

Companies	Percent of sales in dairy products		
	1954	1964	1975
Kraft, Borden, Beatrice, Carnation, Pet	81	58	39
Fairmont, Foremost-McKesson, National Industries (Hawthorn-Mellody)	86	82	23
Total	82	67	36

The regional companies as a group grew at much the same rate as the national companies (the growth rates for both groups are, of course, strongly affected by changing numbers of firms in each group) from 1934 to 1964. Thereafter, some of the regional firms were the beneficiaries of the FTC policy of fostering growth of the "second-tier" companies. Their share of the national market remained at 5 percent from 1934 to 1964, then rose to 8 percent in 1970, dropping to 3.5 percent in 1979 with the decline in firms in the category.

The regional companies have also been diversifying in the last two decades, but not to the same extent as the national companies. The regionals are still primarily food firms, with at least a majority of their sales in fluid milk products.

Cooperatives. Fluid milk processing was not a major activity of cooperatives until recently. In the past 20 years, the number of cooperatives packaging fluid milk has declined sharply with mergers, but the share of the national market has increased from 8 percent to 15 percent.

Integration by Retailers. The first fluid milk plants operated by retail grocery store chains date back to the late twenties and early thirties, when Safeway, Kroger, and Ralphs started. No other grocery chains--which became supermarket chains in the late thirties and forties--joined the ranks of integrators into fluid milk processing until after World War II. After 1950, the pace picked up...

	Companies	Plants
1930	1	1
1935	3	6
1940	3	7
1945	3	7
1948	4	10
1950	6	12
1955	11	20
1960	18	30
1964	21	36
1970	25	47
1980	19	50

This includes only those firms and plants where the principal outlet was to stores owned or controlled by the firm. They include corporate chains, voluntary groups sponsored by wholesale grocers, and retailer cooperatives. In addition, several other supermarket companies operate milk plants which supply other outlets in addition to their own stores. Usually, these are plants which were purchased by supermarket companies as going businesses and the new owners saw no reason to get rid of the outside customers of the former owners. Sales by such companies to their own stores in 1969 accounted for 0.5 percent of sales of commercial processors.

Until recent years, integrated supermarkets accounted for only a small part of total fluid milk processing. In 1964, 3 percent of the sales of all commercial processors were by integrated supermarkets (table 28). By 1970, their share had risen to 8 percent and, by 1980, to 14.2 percent, as new plants were opened primarily by additional companies entering the field. The importance of integrated supermarket plants varies substantially among regions. They started in California. In 1979, the greatest concentration was in California and the Mountain region (figure 2). The next-highest concentration was in the Middle Atlantic and Cornbelt--also early areas of development.

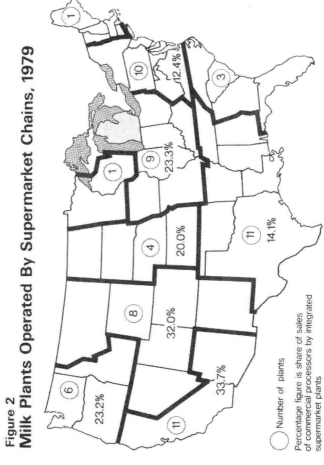

Figure 2
Milk Plants Operated By Supermarket Chains, 1979

○ Number of plants

Percentage figure is share of sales
of commercial processors by integrated
supermarket plants

One stimulus to integration into fluid milk processing by retail food chains was provided by the existence of resale price control under State regulation, as seen most clearly in California. With guaranteed margins for processors, there was often an opportunity for supermarket operators to earn a substantial return on investment in fluid milk processing, since they provided their own guaranteed market. Some operators produced only a very limited line of products and containers, which helped to keep their costs at very low levels, and purchased specialty products and low-volume container sizes from other companies.

Vigorous enforcement of the Robinson-Patman Act against large milk distributors in their contracts with chains also provided some incentive to integrate. Large distributors were the only ones who could handle the business of a substantial retail chain, but the FTC took a dim view of contracts which recognized that the business of a chain could affect the volume of a plant so as to change costs and profitable prices significantly.

Increased incentive in the sixties and seventies was also provided by the slowing of the supermarket boom. By the early sixties, all but the fastest-growing markets in portions of the sunbelt were well provided with supermarkets. Growth could no longer be obtained by displacing small grocery stores. Supermarket organizations looked for other opportunities for growth. Other forms of retailing, especially drug stores and general merchandise stores, were tried by many chains. Vertical growth was also attractive to some, mostly larger chains. Fluid milk processing was a significant growth area for many. Earnings were guaranteed because the firm provided its own market and comparisons were not made to see whether more money could have been made if milk had been purchased from existing processors.

Integrated dairy store chains have also become significant sellers of milk in many markets. Some of this is forward integration by processors seeking volume and a probably smaller part represents backward integration by store operators. Many, but not all, operate their own fluid milk plants. Complete figures are not available for all such plants in the United States, but figures are available for a number of integrated dairy store operations regulated under Federal orders. Sales of these plants increased 73 percent between 1964 and 1971, from 1.3 to 2.2 percent of sales of all fluid milk plants.

Most convenience foodstores purchase fluid milk products from processing plants owned by dairy companies. However, Southland also operates a number of fluid milk bottling plants. These plants sell not only to the convenience stores operated by the company but also to other outlets. In 1969, fluid milk plants operated by integrated dairy and convenience stores combined accounted for 4.2 percent of the sales of commercial processors. In 1980, integrated plants of companies operating dairy and convenience stores combined accounted for 9.6 percent of U.S. sales (Lough, 1981, p. 10).

Market Shares in Local Milk Markets. The exit of more than half of the fluid milk plants operating in 1950 had, by 1965, increased concentration in most smaller milk markets, although it rose relatively little in the larger markets (table 30). Concentration, as

Table 30--Average market share of 4 largest firms, by market size 1/

Size of market	1950	1953	1956	1959	1962	1965
			Percent				
Small	66.7	71.4	77.6	81.0	89.1	88.1
Medium-small	56.9	64.2	66.9	77.4	79.8	82.8
Medium	56.7	63.3	65.1	62.8	66.7	76.0
Large	55.9	57.1	58.1	57.0	57.6	63.3
Very large	45.3	43.4	43.9	41.0	42.8	47.8

1/ Calculated as a link-and-chain index from available data for 72 markets (mostly Federal orders but excluding New York-New Jersey) in 1965 and smaller numbers in earlier years. March of each year, except December of 1965.

measured by the share of the three or four largest processors, has been fairly high in most fluid milk markets ever since they began to develop in the last century. (See table 6 for market shares in the thirties.)

Between 1965 and 1979, the market shares of the four largest firms continued to increase in smaller markets and held steady in very large markets (table 31). In most of the very small markets, four or fewer handlers sold all the milk.

But fluid milk markets grew rapidly in the postwar period, so that the markets in which competition took place (thus, those in which we are interested) were not the same from year to year. This complicates the problem of measurement. Since nearly all of our data comes from Federal order markets, the definitions of those markets for regulatory purposes are the only ones that we have. Federal order markets have grown much larger and fewer in number over the years, but individual markets grew in discrete jumps. Thus, market definitions are always somewhat out of step with economic reality... some markets, recently redefined, correspond rather well with the area in which competition takes place. Others, defined several years ago, are too small.

In recognition of these problems, recent ERS studies of the structure of fluid milk markets (Manchester, 1974a) defined local fluid milk markets without regard to regulatory areas such as Federal orders or political subdivisions. The basic concept used in defining local fluid milk markets was that of "potential competitors." The starting point in each case was a city or metropolitan area. Each fluid milk bottling plant which was a potential competitor in that city or metropolitan area was included in the definition of the local market. This means, of course, that some plants were included as potential competitors in several different markets.

Table 31--Average market share of the 4 largest firms in Federal
order markets, by market size 1/

Size of market	All markets			Comparable markets			
	Markets	1976	1979	Markets	1965	1976	1979
	Number	- -Percent- -		Number	- - - -Percent- - -		
Small	14	95.9	99.6	11	88.1	98.0	99.7
Medium-small	9	83.0	88.8	7	72.2	83.0	86.1
Medium	11	70.4	75.2	9	59.3	72.9	75.8
Large	7	52.2	56.7	5	41.7	55.1	60.0
Very large	5	43.6	41.8	4	38.3	40.8	39.0

1/ Not strictly comparable to table 30, because of changes in numbers and sizes of Federal order markets.

Sources: 1965: Manchester, 1968a and 1968b. 1976-79: Rourke, 1982.

Two major factors determine whether or not a plant is a potential competitor in a metropolitan area: the size of the plant and
the distance from the market center. For this purpose, all plants
located within 50 miles of the market center were included. Plants
located between 51 and 250 miles from the market center were included
if they had monthly sales of 1 million pounds of fluid milk products
or more. By including all plants within 250 miles with sales of at
least 1 million pounds per month as potential competitors, substantially all of the plants that were actually supplying retail stores
in 56 markets in 1969-71 would have been included. In these 56 markets, 91 percent of the milk came from plants within 150 miles of the
market and none from plants more than 250 miles from the market
(Jones, 1973, p. 4). Two-thirds of the milk came from plants within
50 miles and only 1.4 percent from plants between 200 and 250 miles
from the market. Store brand milk produced in integrated plants
accounted for all of the milk which moved more than 200 miles. In a
few isolated instances in the West, which was not included in the
56-market survey, milk produced in integrated supermarket plants
moved as far as 400 miles.

In the case of a firm which operated more than one plant within
the area of potential competition, only the one largest plant was
included. Usually, this was the plant nearest the market center, but
not always.

For plants more than 50 miles from the market center, the total
sales figure was reduced. A large plant located 200 miles from the
market would incur greater costs in delivering milk to the market

than would one within the market itself. Thus, a potential competitor would be at some disadvantage in competing for sales to outlets at such a distance. To allow for this, the actual sales of such plants were reduced by a factor which was proportionate to the cost of delivering milk from the plant to the market. This factor was developed by taking the inverse of the cost function for long-distance shipment of packaged fluid milk products (Moede, 1971) and adjusting the resulting function so that the factor equaled 1.0 at a distance of 50 miles and 0 at a distance of 251 miles. Since the market power of a potential competitor is taken as being proportionate to its sales volume, this procedure made the market power of distant plants proportionate to the cost of moving the milk from the plant to the market and the volume.

The reader should note that figures calculated as described above are not comparable to those calculated on other bases (Manchester, 1964a, 1964b and 1968a; National Comm. on Food Mktg., 1966), including those in tables 30 and 31.

The 144 markets studied in 1969-70 ranged in size from those capable of supplying 3 million pounds per month to those which could supply 466 million pounds per month. The eight markets in the smallest size class averaged 10 plants per market (table 32). The larger markets had more and larger plants. Average concentration (the market share of the four largest firms) declined fairly steadily from 80 percent in the smallest markets to 23 percent in the 14 largest markets. In markets with total sales of less than 75 million pounds per month, the share of national and regional firms varied irregularly from 35 to 43 percent. In larger markets, the share of national and regional firms declined, reaching an average of 20 percent in the largest markets.

Concentration is affected both by size of market and degree of isolation (table 33). Among all sizes of markets, concentration averages highest in isolated markets. In compact markets, concentration averaged from 8 to 24 percentage points lower than in isolated markets of the same size. Dispersed markets--those which are fairly near some other large market--had lower concentration ratios than either of the other two groups.

PLANTS MANUFACTURING DAIRY PRODUCTS

Plants making manufactured dairy products--butter, cheese, dry milk, canned milk, and frozen desserts--have been subject to the same broad economic forces as fluid milk processing plants. Changes in technology on the farm, in transportation, and within the plants have combined to shift the economies of size curve increasingly against the small plant. Plant numbers have dropped and average size increased for many decades (tables 34-36). In Wisconsin, for instance, numbers of butter plants peaked in 1910 and of cheese plants in 1922.

Plants were highly specialized before World War II, producing only one product--butter, American cheese, Swiss cheese, ice cream, or canned milk. In the late thirties and early forties, the advantages of flexible, diversified, multiple-product plants in some situations became apparent. Such plants could shift among products

Table 32--Structural characteristics of 144 fluid milk
markets, 1969-70

Market size	Number of markets	Average market size	Average number of plants	Average market share of--	
				Four largest firms	National and regional firms
Mil. lb. per mo.		Million pounds		Percent	
3-9	8	6	10	79.9	38.1
10-19	13	15	16	72.2	43.4
20-29	19	25	22	60.8	35.6
30-49	23	38	29	53.3	41.9
50-74	19	57	47	46.7	38.3
75-99	14	87	58	39.5	30.9
100-199	34	148	100	30.1	26.9
200-475	14	315	137	23.3	20.4
All markets	144	93	58	46.9	33.8

Table 33--Concentration by market size and isolation, 1969-70

Market size, mil. lbs. per mo.	Average market share of 4 largest firms in--			
	Isolated markets 1/	Compact markets 2/	Dispersed markets 3/	All markets
	Percent			
3-9	96	72	67	80
10-19	85	72	63	72
20-29	4/87	65	40	61
39-49	63	55	38	53
50-74	4/68	51	36	47
75-99	--	43	26	40
100-199	--	34	23	30
200-450	--	24	19	23

1/ Markets where there was no significant number of plants outside
the immediate area and within 250 miles.

2/ Weighted average distance of plants from market center less than
100 miles.

3/ Weighted average distance of plants from market center 100-199
miles.

4/ One market.

78

Table 34-- Number of plants producing manufactured dairy products and average output per plant

Product	1937	1944	1948	1952	1957	1961	1965	1970	1981
				– – – Number of plants – – –					
Butter	4,660	4,022	3,224	2,724	2,062	1,516	1,152	622	238
American cheese	2,312	2,144	1,749	--	1,194	1,023	--	--	481
All cheese	--	2,856	2,295	1,954	1,603	1,410	1,209	963	725
Canned milk	145	144	137	117	85	72	59	--	19
Cottage cheese curd	--	1,652	1,491	1,795	1,609	1,207	918	593	264
Ice cream:									
Regular plants	1/5,003	--	--	--	3,395	--	--	--	--
Wholesale	--	3,656	--	3,101	3,759	--	--	--	--
Large 2/	--	--	--	--	--	1,905	1,560	*1,140	895
Nonfat dry milk	--	498	435	435	456	431	372	219	114
All products 3/	--	9,739	--	--	--	6,134	*5,382	3,749	2,187
				– – – Average production per plant 4/ – – –					
Butter	357	--	375	--	685	--	1,148	1,836	5,160
American cheese	213	375	486	--	853	--	1,339	2,128	5,423
Canned milk	13,121	--	24,693	--	28,796	--	28,694	27,572	39,890
Ice cream	1/54	--	--	--	179	--	464	454	930

--Not available.
* Estimated.
1/ 1938.
2/ Annual output of 20,000 gallons or more.
3/ All plants producing one or more manufactured dairy product--not a total of the above figures since many plants produce two or more products.
4/ Thousand pounds except ice cream in thousand gallons.
Source: Bormuth and Jones, 1959; Carley and Cryer, 1964; Crop Reporting Board.

Table 35--Size distribution of manufactured dairy products plants, 1957-1977

Product and year	Number of plants with annual production (million pounds) of--						
	0-.9	1.0-1.9	2.0-2.9	3.0-3.9	4.0-4.9	5.0 or more	Total
Butter							
1957	1,665	234	82	43	48	*	2,062
1963	935	167	89	60	70	*	1,321
1972	303	51	18	22	14	67	475
1977	191	39	*	82	*	*	312
Cheese							
1957	1,226	233	144	*	*	*	1,603
1963	828	292	179	*	*	*	1,282
1972	375	193	114	96	*	123	901
1977	270	192	82	84	*	165	793
Creamed cottage cheese							
1957	1,500	154	*	*	*	*	1,654
1963	902	184	*	*	*	*	1,086
1972	281	69	40	11	24	57	482
1977	146	54	48	*	65	*	313
Nonfat dry milk (human food)							
1957	143	95	*	112	*	106	456
1963	97	78	*	87	*	143	405
1972	37	26	*	28	*	89	180
1977	39	*	*	1/19	*	74	132

* Included with next-smaller group.
1/ 2,500-4,999

Source: Bormuth and Jones, 1959; Bormuth and Cryer, 1965; U.S. Econ. and Stat. Serv., 1979; U.S. Stat. Rptg. Serv., 1974.

Table 36--Plants making natural cheese, by type of firm, 1958
and 1973

Type of firm	:	Plants		: Percent of : value of : shipments, : 1958	: Percent of : quantity : produced, : 1973
	:	1958	: 1973		
Corporations:	:				
Multi-unit	:	262	137	44.4	25.9
Single unit	:	210	322	15.1	35.1
Cooperatives:	:				
Multi-unit	:	48	55	3.5	18.8
Single unit	:	199	138	10.4	14.4
Individual proprietors	:				
and partnerships	:	638	213	26.1	7.8
Other	:	18	--	.5	--
Total	:	1,373	865	100.0	100.0

Sources: Lough, 1975, p. 24; National Comm. on Food Mktg., 1966,
p. 338.

in response to changing price and profit relationships. By 1944,
there were about 100 such plants out of the total of 9,739 dairy
manufacturing plants (Cowden and Trelogan, 1948, p. 9). By 1961,
515 plants (8 percent of all plants) were fully diversified. Single-
product plants declined from 72 percent of all plants to 44 percent
(Carley and Cryer, 1964, p. 6). The pace of plant diversification
picked up.

The technology of manufactured dairy products production changed
drastically during the postwar period. A general characterization
would be that continuous processes replaced batch processes. The
continuous churn for butter has largely replaced batch manufacture
and printing (molding and packaging consumer units) became a part of
the continuous process rather than a separate operation mostly conducted in other plants by other firms. Together with comparable
processess for other manufactured dairy products, such changes in
technology transformed the manufactured dairy products industries
from many small plants to a relatively small number of large plants.

Changes in demand also contributed to the changing structure of
manufactured dairy products industries. With increasing affluence,
consumers turned from canned evaporated and condensed milk to other
dairy and nondairy products. Ice cream and other frozen dairy products, especially ice milk, enjoyed a surge in demand in the fifties
and sixties but it has now leveled off. Butter was replaced by
lower-priced margarine in many markets. Cheese demand boomed in the
sixties and seventies, especially for Italian cheeses (for pizza and
other pasta dishes) and for American cheese. Nonfat dry milk enjoyed
a boom in the fifties but more recently demand has tapered off.

The output of processed cheese and related products has grown only slightly more than the production of hard cheeses other than Italian. In 1939, the output of these products was about 55 percent as large as that of natural cheese (excluding Italian, cream and Neufchatel) and in 1981 it was 58 percent as large.

Concentration

Despite all of the changes in plant numbers and sizes in the manufacture of dairy products, there are suprisingly few changes in the shares of large firms in individual industries (tables 37 and 38). The two tables present data on the industry basis and the product basis. While the product basis is generally preferred by analysts, data are not available for as long a time span on that basis, so both are shown here. The industry basis takes all of the products of establishments (plants) classified in a given industry--say, butter--as the unit of measurement. The product basis measures only butter, regardless of whether the plants in which it is made are classified primarily in the butter industry or elsewhere, say in fluid milk.

In general, one would conclude that there has been relatively little change in concentration between 1935 and 1972 on the basis of these figures, except for the case of butter between 1967 and 1972. The Bureau of the Census reports dramatic increases in the share of the four largest companies in that five-year span. The change which is reflected in these figures is the many mergers of small cooperatives into larger cooperatives in this period. The biggest element is undoubtedly the merging of the local creameries that were members of Land O'Lakes, Inc., into the parent organization. Thus, the hundreds of individual cooperatives which made butter in 1967 and sold it through the Land O'Lakes sales organization had become part of the parent organization by 1972 and were counted as one company. If the figures were calculated on the basis of sales rather than production, much less change would have been recorded.

In addition to merging member cooperatives into Land O'Lakes, mergers of formerly independent cooperatives into other large regional cooperatives also helped to raise the share of the four largest firms. Another significant factor is that proprietary firms have increasingly withdrawn from the manufacture of butter. Since butter and powder are the residual products in milk utilization and they have been substantially less profitable than cheese, proprietary firms have been increasingly disinterested in their production.

Cooperative Role

Dairy cooperatives have played a part in the processing of butter, cheese, and dry milk products for many years. Nearly all butter and cheese plants in the thirties were local operations. Butter plants received farm-separated cream and cheese plants manufacturing grade milk from a group of producers within a few miles of the plant. Cooperatives accounted for nearly 40 percent of butter output in 1936 and one-fourth of cheese production (table 39).

Table 37--Concentration ratios on the industry basis--Share of industry shipments accounted for by largest companies, 1935-1972

	:	Percent of value of shipments by--			
Industry and year	:	4 largest companies	8 largest companies	20 largest companies	50 largest companies
Butter:	:				
1972	:	45	58	78	92
1967	:	15	22	36	60
1963	:	11	19	31	48
1958	:	11	18	28	45
1954	:	16	24	34	--
1947	:	18	24	32	--
1935	:	17	26	--	--
Natural cheese:	:				
1954	:	25	30	39	--
1947	:	27	32	40	--
Natural and processed cheese:	:				
1972	:	42	53	65	77
1967	:	44	51	61	72
1963	:	44	51	59	69
Cheese:	:				
1935	:	19	23	--	--
Condensed and evaporated milk:	:				
1972	:	39	58	76	92
1967	:	41	56	74	90
1963	:	40	53	71	90
1935	:	45	63	--	--
Concentrated milk:	:				
1954	:	55	68	80	--
1947	:	50	63	76	--
Ice cream and frozen desserts:	:				
1972	:	29	40	58	75
1967	:	33	43	60	73
1963	:	37	48	64	74
1958	:	38	48	59	69
1954	:	36	45	57	--
1947	:	40	48	57	--
Ice cream:	:				
1935	:	33	38	--	--

1935 and 1947 figures are not comparable with other years.
--Not available.

Sources: U.S. National Resources Comm., 1939; U.S. Bu. of Census, 1957, 1966, and 1975a.

Table 38--Share of value of shipments of each class of product accounted for by
the 4, 8, 20, and 50 largest companies, 1954-1972

Class of product and year		Percent of value of shipments accounted for by--			
		4 largest companies	8 largest companies	20 largest companies	50 largest companies
Condensed and evaporated milk	1972	34	52	66	80
	1967	35	47	61	73
	1963	33	42	55	70
	1958	38	48	58	72
Dry milk products	1972	45	59	73	90
	1967	35	45	57	73
	1963	22	30	47	65
	1958	22	33	49	69
Canned milk products (consumer type cans)	1972	69	83	98	100
	1967	62	81	95	100
	1963	66	78	93	100
	1958	78	85	97	99
	1954	79	86	97	(NA)
Concentrated milk, shipped in bulk	1972	29	47	75	98
	1967	31	49	75	94
	1963	41	54	74	91
	1958	38	54	76	92
	1954	45	58	73	(NA)
Ice cream mix and ice milk mix	1972	16	30	50	72
	1967	15	26	44	70
	1963	17	26	45	69
	1958	23	35	56	77
	1954	24	33	46	(NA)
Ice cream and ices	1972	27	37	54	70
	1967	32	42	57	68
	1963	34	43	57	65
	1958	35	44	54	62
	1954	33	41	52	(NA)
Creamery butter	1972	37	47	61	78
	1967	14	20	33	51
	1963	8	14	25	40
	1958	11	15	24	37
	1954	14	19	28	(NA)
Cheese, natural and processed	1972	40	51	62	74
	1967	45	53	62	71
	1963	45	50	59	68
Natural cheese, except cottage cheese	1972	36	46	63	77
	1967	38	44	55	68
	1963	34	39	49	61
Process cheese and related products	1972	60	74	86	95
	1967	72	84	94	99
	1963	67	74	(NA)	(NA)

Source: U.S. Bu. of the Census, 1957, 1966, and 1975a.

Table 39—Cooperatives manufacturing or distributing dairy products

Product	1936	1944	1957	1964	1973	1981*
			Number of cooperatives			
Butter	1,444	1,164	888	734	207	--
Natural cheese	562	501	323	294	187	--
American cheese	--	--	248	--	--	--
Other cheese	--	--	75	--	--	--
Nonfat dry milk	139	--	191	212	57	--
Cottage cheese	14	--	108	126	64	--
Ice cream and ice milk	--	--	130	143	60	--
All manufactured products	--	--	1,180	856	291	--
			Cooperatives' share of total output (percent)			
Butter	39	*55	58	65	66	75
Natural cheese	25	16	18	21	35	50
American cheese	--	--	23	--	--	--
Other cheese	--	--	8	--	--	--
Dry milk products	17	56	57	72	85	90
Cottage cheese	1	--	7	15	13	14
Ice cream and ice milk	1	--	4	5	5	5

* Estimated.
-- Not available.

Sources: Black, 1935; U.S. Bu. of Ag. Econ., 1950; Gessner, 1959; Herrman and Welden, 1941; Hirsch, 1951; Metzger, 1930; Tucker et al, 1977.

During World War II, the government made strenuous efforts to obtain the utilization for human food of the skim milk which had been kept on the farm and fed to animals when farm-separated cream went to the creamery. Sixteen milk drying plants were built by the Government using lend-lease funds and operated largely by cooperatives which subsequently acquired the plants. In addition, milk drying equipment was installed in 9 existing cooperative plants. The cooperatives' share of the output of dry milk products increased from 17 percent in 1939 to 56 percent in 1944.

Cooperatives made about one-third of all cheese in 1926. Proprietary firms increased output much more rapidly than cooperatives in the next 20 years and the cooperative share declined by about a half, despite a 20-percent increase in output from cooperative plants.

Cooperative butter output went up, along with dry milk products, during World War II. In 1943, cooperatives made 70 percent of the butter produced in Minnesota, 59 percent in Wisconsin, and 54 percent in Iowa. These three states produced 40 percent of U.S. butter.

Cooperatives maintained approximately constant market shares through the late forties and the fifties. By 1964, nearly two-thirds of all butter was produced by cooperatives, a share that has grown to about three-fourths since then. Dry milk products now come largely from cooperative plants. A major reason for cooperative dominance in butter-powder production is the fact that these are the residual products of the entire dairy system. Volumes vary widely both over the dairy cycle and within the year and they have become less and less attractive to proprietary concerns seeking profitable investments.

In the late sixties and seventies, cooperatives more than doubled their share of natural cheese production. This is partly a result of greater profitability in cheese than in butter-powder, as demand for cheese has grown rapidly. The regional cooperatives organized in the late sixties consolidated manufacturing plants and built new ones. Many are either cheese plants or diversified plants that can produce either butter-powder or cheese, depending upon relative returns.

IN SUMMARY

Fluid milk processing has changed from a business of thousands of plants to one of large plants since the thirties, in response to shifts in economies of size in plant operation and distribution. Producer-distributors are nearly gone. Commercial processors now number about a thousand, about a tenth of the numbers in the thirties. The average plant processes 25 times as much milk.

National firms lost market share from the thirties until about 1950, then gained with many mergers in the fifties and early sixties, and lost share in the seventies until two additional companies grew to national stature. Cooperatives have grown steadily since the thirties in fluid milk processing. The big gainer was integrated supermarket chains which now have 13-17 percent of the market. The national companies have diversified broadly and are much less specialized in dairy products than they used to be.

Competition in fluid milk takes place in the local market. These markets have grown in size. In the smallest markets, all of the business is now in the hands of four processors or less, but in the largest markets concentration is now lower than it was thirty years ago.

Manufactured dairy products plants have also declined in numbers and increased in size. The large dairy companies have withdrawn from the manufacture of dairy products to some extent and coooperatives have become increasingly important in butter, nonfat dry milk, and cheese.

7
Changes in Marketing and Consumption

MARKETING PACKAGED MILK

The earlier section discussed the changes in the number and type of fluid milk processing plants. The decline of producer-distributors, the waxing and waning of national dairy firms, and the growth of integrated supermarket chains had their impacts on packaged milk marketing. However, there were other firms in packaged milk markets.

There was a tremendous growth in the number of subdealers in the thirties. These were independent dealers who purchased bottled milk from processors and delivered it to homes or stores. The Depression growth was due largely to a desire on the part of the unemployed to find some kind of work. Some were former processors who could no longer afford to operate a bottling plant or deliverymen for processors who had been laid off.

In New York City in 1939, approximately 4 percent of all milk was sold by subdealers or 10 percent of all home-delivered milk (Spencer et al, 1941). Their business was concentrated in middle-income areas with relatively few stores, especially in Brooklyn and Queens. About half were former employees of milk distributors. Under regulations of the New York City Department of Health, subdealers had been licensed for the first time in July 1939, although many had been masquerading as hired deliverymen for processors for several years during the Depression. The subdealer business in New York State burgeoned thereafter. By November 1966, over a third of fluid milk sales in New York State were made by subdealers. About 90 percent of the subdealer business was in the New York Metropolitan area. There were 1,180 subdealers throughout the State operating through the year, with 75 percent in the Metropolitan area and another 15 percent in the suburban area to the north of the city. Some subdealers sold over 1 million pounds of fluid milk products per year. A major part of the growth of subdealers in New York State is attributable to restrictive licensing by the State (Lasley, 1972).

Nearly 45 percent of the sales of fluid milk products in the New York-New Jersey marketing area in November 1981 were by subdealers. They made most of the home delivery sales, although home delivery accounted for less than 1 percent of total sales (Market Admin., N.Y.-N.J., 1982).

The Connecticut milk business was divided 51 percent to commercial processors, 39 percent to producer-distributors, and 10 percent to subdealers in 1940 (Bressler, 1952, p. 30). But the subdealers were nearly as numerous as the commercial processors--258 subdealers and 304 commercial processors. In the next quarter century, numbers of commercial processors dropped more than 70 percent, but subdealer numbers declined only 27 percent.

In two other States for which we have license data--Massachusetts and Pennsylvania--numbers of subdealers peaked in the early forties, dropped off a bit in the late fifties, and rose again to the levels of the late thirties by 1965.

In California, data on the share of total sales by subdealers have been available since 1951. In the early fifties, subdealers' share rose to nearly 12 percent of total sales, then dropped to a low of about 8.5 percent in the late fifties, and rose gradually to 11 percent by 1969. The rise in the seventies was much more rapid-- to 14 percent in 1971 and to 16 percent in 1976.

In Chicago, the subdealer business burgeoned during the thirties, in large part because of strong union opposition to store sales and the union's ability to maintain relatively high wage rates. In 1941-42, when the Bureau of Labor Statistics made a study, earnings of subdealers were modestly below those of union deliverymen and the subdealers were more heavily concentrated in the store business (Gresham and Block, 1942). Subdealers had existed in Chicago all through the twenties, but they had not threatened the wage or price structure of the unionized plants until the Depression. The Chicago union wage scale set in 1927 was the highest in the country at $50 per week for retail routemen.

In 1932, Meadowmoor Dairies entered the Chicago market to distribute milk exclusively through vendors, buying directly from farmers at a flat price, cutting prices to stores. This led to a protracted war in the Chicago market, including the bombing of the Meadowmoor plant and numerous lawsuits (Fortune, 1939a).

As smaller fluid milk processing plants closed or sold out to larger firms in the fifties and sixties, the number of subdealers and their share of the total business increased. Many of the smaller firms closed their plants and became subdealers for larger firms, often the firm which purchased the plant. Between 1945 and 1957, 16 percent of the 426 firms leaving the Federal order markets for which we have information became subdealers. In 1958-65, information on the circumstances surrounding exit is more complete and one-third of all exits from fluid milk processing became subdealers.

Estimates based on the Censuses of Business and Manufactures indicate that there were about 8,400 subdealers and distribution branches in 1958 and 1963, with the numbers declining in later years (table 40). An independent estimate made for the National Commission on Food Marketing by the author for December 1964 found--

7,720 subdealers
<u>1,315</u> distribution branches
9,035 total

Table 40--Milk subdealers and distribution branches 1/

Item	1954	1958	1963	1967	1972
	-- -- -- -- -- -- -- -- -Number- -- -- -- -- -- -- -- --				
Establishment:					
Retail	5,450	5,870	5,800	5,470	5,110
Wholesale	1,200	1,930	1,940	1,800	1,890
Sales branches	270	610	693	600	600
Total	6,920	8,410	8,433	7,870	7,600
	-- -- -- -- -- -- -Million dollars- -- -- -- -- --				
Sales of milk and cream :					
Retail	350	400	400	390	365
Wholesale	315	720	760	1,200	2,100
Sales branches	210	490	645	560	850
Total	870	1,610	1,805	2,150	3,315
	-- -- -- -- -- -- -Million pounds- -- -- -- -- -- --				
Sales of milk and cream :					
Retail	3,090	3,215	3,180	2,830	2,200
Wholesale	3,150	7,900	8,380	12,240	19,820
Sales branches	2,690	4,590	6,220	5,120	7,330
Total	8,930	15,705	17,780	20,190	29,350
	-- -- -- -- -- -- -- -- -Percent- -- -- -- -- -- -- -- --				
Quantity sold as percent:					
of sales by commercial :					
processors:					
Subdealers	11.2	19.0	18.1	23.6	31.6
Distribution branches :	10.9	15.4	17.9	15.7	22.4
Total	22.1	34.4	36.0	39.3	54.0

1/ All except the last section are in terms of the Census classification of establishments. Retail dealers make a majority of their
sales to consumers. Wholesale dealers make more sales to stores,
restaurants, and institutional outlets. Sales branches are identified
as outlets owned by fluid milk processors. But many retail and wholesale dealers are also distribution branches of fluid milk processors.
In the last section, estimates of sales by independent subdealers and
by distribution branches owned by processors are presented, based on
Census Enterprise Statistics and an independent count of distribution
branches.

National firms operated 423 distribution branches, regional firms 131, and local proprietary single-unit firms 445. Most were in the Middle Atlantic, East North Central, South Atlantic, and the Pacific regions.

Distribution branches and subdealers handled about 30 percent of the milk in the fifties and early sixties. Thereafter, a sharp rise took place, reaching 40 percent in 1967, and 54 percent in 1972.

The economics of processing and distribution contributed heavily to the growth of both groups. Economies of scale in processing encouraged plant closings by firms owning several plants in the same general area. Many of the closed plants continued to operate as distribution branches, utilizing the cold storage capacity and handling facilities of the former milk plants. Other distribution branches were established at new locations.

Larger subdealers responded to the same economic forces as distribution branches, operating a distribution facility but under independent ownership. Smaller subdealers took delivery either from the processing plant or a distribution branch and distributed to their own customers. Under these arrangements, processors retained the business of many customers, but they were served by independent subdealers. Independent subdealers with no employees or only a few mostly escaped union organizers and could pay lower wages or commissions than the unionized processors. Probably two-thirds of the remaining home delivery sales are made by independent subdealers today.

Changing Outlets

The most far-reaching change--in terms of its effect on milk distribution--was the shift from home delivery to store sales of milk. The significance of this shift lies in the sharply altered market power relationships in packaged fluid milk distribution. In the thirties, with 70 percent of all milk home delivered, fluid milk processors as a group had nearly all of the market power as sellers. While fluid milk processors and distributors were much more numerous then than now, markets were typically composed of a handful of large firms and a much larger competitive fringe of smaller firms. Consumers retained the opportunity of switching from one dealer to another, so there was a strong potential of competition among dealers. The dealers' association or bottle exchange in many markets provided a forum for toning down price competition, although explicit price fixing mostly in contract sales was discovered and prosecuted from time to time by local and Federal agencies. The hunger for volume and profits engendered by the Great Depression probably provided more restraint on price fixing than the prosecutors.

By the late sixties, the transition from home delivery to store sales of milk was completed--not in market share but in terms of dominance. The milk distributors no longer set the pace of competition (or lack of it). That role had been assumed by the supermarket chains. In the following decade, the transition was completed, with

home delivery only for the well-to-do or the very large milk consumer...or in isolated areas where nonunionized labor or independent
subdealers or "bobtailers" dominated home delivery. The story of
these changes provides considerable insight into changes in market
power in milk distribution in the past 50 years.

In the early thirties, nearly three-quarters of all milk for
home use was home delivered. The share declined slowly—perhaps 2
percentage points every five years—until 1950 and then the rate of
decline began to pick up speed (table 41). By 1960, only a bit over
half was home delivered, plummeting to 7 percent today. Only about
half of the home delivered sales are on routes operated by commercial
processors, with the remainder by subdealers and a small part by
producer-distributors.

Table 41—Home delivery of whole milk, 1935-1977

Year	:	Percent of milk home delivered 1/	:	Price differential Home delivery over store	
	:		:	Cents per half gallon 2/	Percent of delivered price 3/
1935	:	72		1.2	5.3
1940	:	70		3.0	13.2
1945	:	67		3.0	10.5
1950	:	55		4.1	10.9
1955	:	49		5.0	11.7
1960	:	44		5.7	12.1
1965	:	38		5.8	12.4
1970	:	27		9.4	16.8
1975	:	13		12.3	16.3
1977	:	8		12.5	15.3

1/ Milk for home use only (excludes restaurants and institutions).
2/ Weighted average for all sizes of containers (quart, half gallon,
and gallon) and all types of whole milk (creamline, homogenized, and
homogenized Vitamin D) in city markets.
3/ Differential as percent of home delivered price.

One part of the incentive for the shift from home delivery to
store purchase of milk was relative prices. In the thirties, the
typical differential between home and store prices—where any existed
—was 1 cent per quart (table 42). In many markets, there was no
differential at all. Indeed, the existence of a differential was a
source of great controversy during the early efforts to control resale
prices—under Federal marketing agreements in 1933 and under State
milk control throughout the thirties. A store differential did not
become legal in a few States until the seventies.

The average differential widened sharply during the late thirties, then stayed fairly constant as a percentage of price until the
mid-sixties. By 1970, largely as a result of increased sales of

Table 42--Store margins and differentials from home delivered
prices, 122 markets, January 1938

Store margin 1/ (cents per qt.)	:	Number of cities with store differentials 2/ of--			
	:	0 cents :	1 cent :	2 cents :	Total
0	:	--	--	4	4
1	:	4	9	2	15
1 1/2	:	17	9	--	26
2	:	60	12	2	74
2 1/2	:	2	--	--	2
3	:	1	--	--	1
Total	:	84	30	8	122

1/ Store margin=store price to consumers less price paid by store.
2/ Store differential=retail home-delivered price per quart less
store price to consumers.
Source: Tinley, 1938, p. 106. Compiled from FLUID MILK MARKET
REPORT, U.S. Bu. of Agric. Econ., Jan. 1938.

larger containers through stores, it widened to about 16 percent of
the home delivered price.

The major shift in outlets for fluid milk products was from home
delivery to store sales over the past fifty years (table 43). Coin-
cidentally, the shares of the two outlets have about reversed, with
three-quarters of all sales home delivered in 1929 and three-fourths
through stores in 1977. The store share was about 8 percent in 1929
and the home-delivered proportion 7 percent in 1977. The types of
stores have changed drastically, with supermarkets--unknown in the
early thirties--accounting for more than half of store sales today.
Dairy and convenience stores increased from 2 percent to about 10
percent over the period.

Innovations

The story of changes in the marketing of packaged milk has many
strands. The shift from home delivery to store sales--and especially
the role of supermarkets--is a major part of the tale. Related developments in containers, products, home delivery practices, retailer
brands, and dairy stores are summarized in terms of the appearance of
each of 14 innovations in 158 city milk markets (table 44). For most
of this period, we do not have figures on the actual proportions of
milk sold in various containers or of the different marketing practices. But we do have fairly good information on the dates when
these innovations were first made in each market.

These data on innovations were first developed in a study of the
relationship between the structure of fluid milk markets, institutional variables, and performance in terms of innovations and marketing margins (Manchester, 1974a). The numbers in the table reflect
the relative size as well as the numbers of milk markets in which

Table 43--Marketing channels for fluid milk products

Outlet	1929	1939	1948	1954	1969	1977
			Percent			
Home delivered	73.3	70.3	56.2	50.0	28.0	6.6
Plant and farm sales	5.9	5.0	2.0	2.0	1.9	1.5
Stores:						
Supermarkets:						
Integrated	0	*	*	1.0	7.1	13.4
Other	0	.5	5.0	11.5	14.9	25.0
Dairy and con-						
venience stores:						
Integrated	*	*	*	.1	4.4	5.6
Other	2.0	2.7	2.4	2.8	3.1	4.2
Other stores:	5.9	6.0	20.6	19.1	21.5	27.1
All stores	7.9	9.2	28.0	34.5	51.0	75.4
Institutional outlets:						
Military messes	0	*	1.1	2.8	1.6	1.2
Schools	*	*	1.3	2.1	6.5	7.1
Restaurants,						
hotels, and						
institutions	12.9	15.5	11.4	8.4	8.6	5.7
All institutions	12.9	15.5	13.8	13.3	16.7	14.1
Other	*	*	*	.2	2.4	2.4
Grand total	100.0	100.0	100.0	100.0	100.0	100.0

* Less than 0.5 percent.

innovations had been made by a given date. The weights for individual markets are constant throughout the period--thus, they do not reflect changes in the relative size of markets over the years. The figures are only for the 158 markets studied, not for the entire United States, but they give a good picture of the relative rate of introduction.

In general--although not universally--innovations were made earlier in large markets than in smaller markets. Thus, the rapid rates of innovation reflected in the tables are partly a reflection of the larger size of the early-innovating markets and partly it reflects the typical S-shaped growth curve.

Pasteurized milk largely predates the period covered here and is not included. Louis Pasteur started his experiments with pasteurization in 1856 and the first commercial milk pasteurizing machines were introduced in 1895. Pasteurization of market milk started in Cincinnati in 1897, in New York, Philadelphia, and St. Louis by 1900, and in Milwaukee, Boston, and Chicago by 1908 (Sommer, 1946, p. 106). The first compulsory pasteurization law took effect in Chicago in 1908, applying to all milk except that from tuberculin-tested cows. By 1924, 98 percent of all milk in 9 large cities was pasteurized, but

Table 44--Innovations in packaged milk marketing, 1930-1970

Innovation	Percent of markets 1/ where innovation had been introduced by--								
	1930	1935	1940	1945	1950	1955	1960	1965	1970
Containers:									
Paper	7	24	64	72	99	100	100	98	100
Half gallon	0	1	34	47	70	95	88	86	86
Gallon	0	3	26	37	41	59	45		
Plastic	0	0	0	0	0	0			
Products:									
Homogenized milk	0	2	67	85	100	100	100	100	100
Half and half	0	0	9	21	57	84	97	100	100
Lowfat milk	0	0	0	0	1	16	43	83	100
Home delivery practices:									
Half gallon or gallon	0	1	30	49	60	84	100	100	100
Quantity discounts	0	0.1	14	31	38	53	66	72	83
Three deliveries per week	0	0	0	6	20	80	82	99	100
Less than three deliveries per week	0	0	0	0	0	0.1	2	30	88
Retail brands:									
From own plant	1	4	7	14	18	33	57	72	87
Custom packaged	1	5	6	6	7	21	56	83	99
Dairy stores	12	19	44	44	56	66	77	84	85

1/ Weighted by milk sales.

most small municipalities did not reply to the survey since they had no milk regulations at all (Fuchs and Frank, 1939, p. 29). In 1936, three-fourths of the milk supply of all municipalities of more than 25,000 population was pasteurized (table 56).

Pasteurization was a public health measure. Seldom was it an innovation in the competitive sense of one processor adopting it in hope of obtaining a sales advantage over his competitors, although this undoubtedly happened at times when a processor had been compelled to start pasteurizing milk for his home market and therefore introduced it into another nearby market where it was not yet required.

The returnable quart glass bottle was universally used by 1930. It had been invented in 1884 and automatic filling and capping equipment patented in 1886. There were many variants of the basic container, in shape, weight, and color (the amber bottle impeded the passage of the sun's rays when the early morning sun hit the bottle on the doorstep). We do not deal with these variants here.

The paper single-service container was patented in 1906. It was introduced commercially in New York City in 1929 by Sheffield Farms for use in the retail store trade (Spencer, 1936). New York City, with a store on most every corner and several in most blocks, was an ideal city to encourage store sales of milk. The customers had only a few steps to go to the store in many cases. Store sales in such densely populated cities did not have to wait upon the widespread ownership of automobiles, as they did elsewhere.

Nevertheless, the original introduction fizzled out after a few years in the retail store trade, although paper containers persisted in the food service business (Spencer, 1936). In 1935, A&P started using paper quarts in its New York City stores and this time they caught on (Dairy Record, 1935). They sold at the same price as glass until State resale price controls were removed in May 1937.

The time span between the invention of the paper container and its commercial introduction was due to two factors...the need for an innovator and the availability of mechanically refrigerated retail cases. The latter came into use in the twenties, opening up the opportunity for an innovator to bring in paper milk containers in retail stores.

Paper containers spread fairly rapidly during the thirties--to markets with a quarter of all sales by 1935 and nearly two-thirds by 1940. By 1950, nearly all markets were using paper containers for part of their milk.

Larger containers began to come in during the late thirties, first the half gallon--in markets with a third of total sales by 1940--and then the gallon--in markets with about a fourth of all sales in 1940, although the share of volume in larger containers grew much more slowly. Both increased fairly rapidly during the war, in part because of the emphasis on economy in packaging and distribution. The half gallon spread to markets with 70 percent of the volume by 1950 and to all markets by 1960. The spread of the gallon was slower, due in part to the unsatisfactory nature of the paper gallon and the weight of the glass gallon container. It was not until the introduction of the plastic gallon in the late fifties and early sixties that it spread to all markets.

Those introducing containers especially suited to the store trade were the mavericks in many markets--either an outsider trying to enter the market or a smaller firm trying to expand its sales. Established firms with a large home delivery business fought back. Half gallons and gallons were introduced on home delivery routes nearly as rapidly as for store sales up until 1945. Quantity discounts were tried in many markets to encourage home delivery sales. For large-volume buyers, home delivery prices might be nearly the same as store prices.

The frequency of home delivery was reduced to lower costs. The change from every-day to every-other-day was decreed by the Federal government during World War II primarily to reduce the use of rubber tires by delivery vehicles. Delivery three-times-per-week (no delivery on Sunday) came in during the early fifties and twice-a-week or two days one week and three the next during the sixties. As we have seen, valiant efforts to stem the switch from home delivery to retail store sales were in vain.

Dairy stores were opened in many markets during the thirties. They typically emphasized lower price and larger quantities, featuring gallon jugs in many instances. After the war, they started up in many other markets, often to an outbreak of price warfare as they upset the established way of doing business.

Retailer (private label) brands were the latest development. A few grocery chains had their own brands in the thirties. The handful of grocery chains with their own fluid milk plants carried their own brands and a few others, but the big increase came in the late fifties and sixties as chains perceived the profit opportunities in a switch from handling fluid milk essentially as a commission item--as they had in most cases until then. The practice had been to carry three or four of the most popular processor brands and to have the processor's deliveryman service the milk case.

Retailers adopted central milk plans, typically contracting with one processor to provide private label fluid milk with reduced service. Frequently, drop deliveries were made to the retailer's platform or store cooler in pallet loads, in contrast to the full service previously prevailing where the routeman had stocked the dairy case and performed merchandising services. This greatly reduced delivery costs and prices reflected the cost reduction. The processor was often permitted to supply his own brand of milk in addition to the private label. In many cases, it sold at a higher price than the supermarket's brand. Since the supermarket brand included only the best-selling fluid milk products, the processor brand was the only one for minor products such as cream, yogurt, and sour cream (Manchester, 1971; Fallert, 1971).

In 56 markets in the Northeast, Midwest, and South surveyed in 1969-71, supermarkets accounted for 87 percent of all whole milk sold by the stores surveyed, with convenience and dairy stores sharing about equally in the remainder (Jones, 1973, p. 9). Store brands represented 57 percent of sales in all stores, processor brands 37 percent, and processor brands sold in dairy and convenience stores owned by the processor 6 percent. Slightly over half the store-brand milk came from plants owned by the retail organization.

Store-brand milk accounted for 65 percent of sales of chain-stores, 34 percent of the sales of voluntary and cooperative groups, and 14 percent of the sales of independent stores (table 45).

Homogenized milk was first sold commercially on a continuous basis by Torrington Creamery Company, Torrington, Connecticut, in 1919. The next recorded introduction was in Ottawa in 1927 and then by McDonald Dairy, Flint, Michigan, in 1932 (Trout et al, 1961). It was widely adopted in the last half of the thirties--by 1940, markets making two-thirds of the sales had some homogenized milk and, by 1950, it was in every market.

Homogenized milk was a natural companion of the paper container, since the customer could no longer see the creamline in the glass bottle. By 1960, it was the dominant form of whole milk and today creamline milk cannot be found in most markets.

Half and half started in the late thirties, got a boost during World War II when the sale of heavy cream was prohibited, and has been generally available for the last 20 years. Nowadays, half and half dominates the much-reduced market for fluid cream products.

Lowfat milk (1 and 2 percent butterfat compared to 3.25 percent or so for whole milk) is a child of the postwar period. Its availability and use grew rapidly in the fifties and sixties. Today, 32 percent of all beverage milk sales are of lowfat milk, compared to 0.1 percent in 1954. Whole milk is down from 92 percent of all beverage milks in 1954 to 58 percent in 1981. Skim milk is not treated as an innovation, since it has been available for many years--its share of beverage milk sales grew from 2 percent in 1954 to 5 percent in 1981.

Growing Size of Packaged Milk Markets

One result of many of the changes in processing and marketing fluid milk was that milk markets became much larger in a geographic sense. Markets that were very local in character and limited to one city or town in the thirties are now much, much larger. Processors and distributors in a given location are in competition with other processors located up to several hundred miles away. A wide variety of technological, economic and institutional changes contributed to the a broadened base of competition in fluid milk markets.

The economics of distribution over wide areas were affected by--

o Truck size and speed.
o Availability of highways and possible speed.
o Highway weight limits.
o Availability and cost of refrigeration equipment on trucks.
o Sanitary regulations.
o Containers for milk, especially their weight.
o Keeping quality of milk.
o Economies of size in processing and distribution.

The introduction of the motortruck in city distribution in the twenties brought about a quantum increase in the area of distribution from a given plant or distribution branch. Later developments in improved highways and faster trucks increased the average speed of

Table 45—Stores selling various brands of whole milk, by type of store and type of market, 56 markets, 1969-71

Brand and packaging arrangement	Chainstores			Independent stores			Voluntary and cooperative stores			Total
	Controlled markets	Uncontrolled markets	Total	Controlled markets	Uncontrolled markets	Total	Controlled markets	Uncontrolled markets	Total	
	Percent 1/									
Stores selling only own brand:										
Integrated	4.2	20.2	24.4	--	--	--	0.4	--	0.4	24.8
Custom-packaged	4.4	2.8	7.2	--	--	--	--	--	--	7.2
Subtotal	8.6	23.0	31.6	--	--	--	.4		.4	32.0
Stores selling own and processor brand:										
Own brand:										
Integrated	.9	7.4	8.3	--	--	--	.1	--	.1	8.4
Custom-packaged	5.3	14.9	20.2	--	0.1	0.1	.1	1.9	2.0	22.3
Processor brand by:										
Processor supplying custom-packaged	2.9	3.7	6.6	2/	2/	2/	.1	1.1	1.2	7.8
Other processor	.5	3.0	3.5	--	--	--	.1	.6	.7	4.2
Subtotal	9.6	29.0	38.6	.1	.1	.1	.4	3.6	4.0	42.7
Stores selling only processor brands:										
One brand	7.9	2.2	10.1	.2	.2	.2	2/	.7	.7	11.0
Two brands	2.6	3.2	5.8	.2	.2	.2	.1	--	.1	6.1
Three brands	2.3	2.1	4.4	2/	2/	2/	2.1	--	2.1	6.5
Four brands	.4	.2	.6	.2	--	.2	2/	--	2/	.8
Five brands	--	--	--	--	--	--	2/	--	2/	2/
Secondary brands	2/	.8	.8	--	--	--	--	--	--	.8
Subtotal	13.2	8.5	21.7	.2	.4	.6	2.3	.7	3.0	25.3
Grand total	31.4	60.5	91.9	.2	.5	.7	3.1	4.3	7.4	100.0

-- means not applicable.
1/ Estimated percentage of milk sold by store groups surveyed, based on facings for within-store weights and total sales of all products by the store group in the market for between-group weights.
2/ Less than 0.05 percent.

Source: Jones, 1973.

trucks on the highway from about 40 miles per hour at the end of World War II to nearly 55 miles per hour (see table 21). While city speeds were undoubtedly much less, the improvements in truck design and the availability of limited-access highways (the beltways around most major cities, for example) made comparable increases in distribution areas possible. Larger truck sizes also made it possible to move much bigger quantities of milk in one load. Highway weight limits, while never a major factor in packaged milk movement, have also increased.

The limitations on milk movement imposed by local sanitary regulations are now largely, although not entirely, gone.

Probably the most important factor--certainly the most visible-- was the change in milk containers. The introduction of the paper milk carton reduced the weight of container and case (not including the milk) from 70 pounds per case of 12 standard glass quarts to less than half that weight in paper containers. While different styles of bottles and cases reduced the weight, it was still substantially more than paper.

Improved keeping quality of milk, due to improved production, handling, and processing methods, made possible less-frequent delivery of milk. Thus, customers could be served less often with larger deliveries. This reduced costs per unit and let processors and distributors reach out farther to look for outlets. The shift from home delivery to store sales also greatly increased the average size of a delivery.

The shift in the plant size curve increased the cost advantages of the large plant compared to the small plant and the large plants could distribute milk at greater distances and still be competitive with smaller plants closer to the outlet.

The end of World War II brought a close to gasoline and tire rationing and resumption of highway construction. This meant that milk dealers' operations were more mobile and so were their potential customers. Many dealers saw the opportunity to expand sales by moving out of the confines of the home town, primarily looking for store accounts to be served with paper containers.

In 1951, Williams reported that trucking packaged milk up to 150 miles was then common in the East and in the Great Plains States from 300 to 400 miles was usual. One plant in Texas was shipping milk up to 600 miles.

By 1952, nearly 500 distributors in the North Central region were distributing milk outside their home town or city, with 23 percent of them shipping milk 100 miles or more. Eighty-three percent of the towns and cities in the region were served with milk from another town or city, averaging 40 miles away. Ninety-one percent of the distributors had begun sales outside their home city or town since 1945 (N.C. Reg. Dairy Mktg. Comm., 1953).

Specialized firms with mobile operations designed to serve retail outlets over large regions developed in the late forties and especially in the early fifties. They operated fleets of large refrigerated trailers to move milk from the plant to central distribution points, sometimes hundreds of miles away. Some distribution points were simply parking lots for the trailers, where the milk was transferred to smaller trucks for distribution to stores and, sometimes, to homes.

Half-gallon paper containers were essential for these operations (Harris, 1966, pp. 10-11).

In the thirties, packaged milk generally moved 50 miles or less from plant to outlet. By 1950, it was moving 100-125 miles and occasionally more (Spencer and Christensen, 1955, p. 41). As transportation improved and barriers became less important in the fifties and sixties, the distances gradually increased to about 200 miles by 1970 (table 46). Rising fuel prices since 1973 have limited further growth.

Table 46--Proportion of whole milk sold, by distance from processor to store, type of brand, and type of market, 56 markets, 1969-71

Type of brand and market	Distance in miles from processor to store 1/					Total
	0-49.9	50-99.9	100-149.9	150-199.9	200-249.9	
	--------	--------	--------Percent of milk--------			
Processor brand:						
Controlled markets	2.9	4.0	.4	--	--	7.3
Uncontrolled markets	37.4	3.5	1.8	2.8	--	45.5
Subtotal	40.3	7.5	2.2	2.8	--	52.8
Store brand:						
Integrated:						
Controlled markets	6.0	2.3	.2	.9	.4	9.8
Uncontrolled markets	8.3	3.3	2.4	3.7	1.0	18.7
Subtotal	14.3	5.6	2.6	4.6	1.4	28.5
Custom packaged:						
Controlled markets	3.9	1.5	.1	--	--	5.5
Uncontrolled markets	8.8	3.2	.5	.7	--	13.2
Subtotal	12.7	4.7	.6	.7	--	18.7
Total store brand	27.0	10.3	3.2	5.3	1.4	47.2
Total	67.3	17.8	5.4	8.1	1.4	100.0

--means not applicable.

1/ Distance outside city limits.

Source: Jones, 1973, p. 4.

Competition in Packaged Milk Markets 1/

The broad outlines of the competitive process in city milk markets from the twenties up until about 1960--the period of what might be called "dealer dominance"--have been described by Harris (1966, pp. 6-9). In the larger markets, a few large firms competed as oligopolists. Small firms normally conformed to the price and sales policies of the oligopolists but had nothing to do with the determination of these policies. They were effectively barred from direct competition in phases of the market dominated by the large firms. The place of the small dealer in a city market was usually somewhat precarious. Typically, he sought out a place on the fringes or within the crevices of the market. Some specialized in suburban business in the vicinity of their plants. A housing development might become the object of intensive sales effort. Owners of small groceries or small eating places might be sought as customers. Some started dairy stores, selling milk to consumers in larger-size containers at lower prices.

The larger dealers operated on a broader competitive base and, in general, were more firmly entrenched in the market. They delivered to homes and stores throughout the market. They were able to obtain and hold the business of the larger stores, restaurants, and institutions. The largest dealers had the resources to offer the greatest concessions in special discounts, equipment, or credit when these concessions were necessary to win or protect accounts from their smaller adversaries. The struggle for individual accounts always went on behind the peaceful facade of the going price structure. Just which accounts were the points of contention at any time depended on the initiative of dealers seeking to extend their own sales, to curtail sales of others, or to contest for new customers who did not clearly fall within the province of any single dealer.

Each dealer usually operated within a certain sphere of competitive influence. Even the larger dealers normally respected the place in the market that the smallest dealer had made for himself. This mutual, tacit respect was sometimes expressed as "Live and Let Live" policy. In part, it was a recognition that direct, unfettered competition had destructive potentials for all participants and, in part, it was an acceptance of the fact that each dealer's home ground was his position of strength. The small dealer with customers concentrated near his suburban plant was not easily displaced by a larger dealer. On the other hand, the chain supermarket served by a large dealer was not a potential customer for smaller dealers.

The competitive situation was, however, always fluid. The areas of influence were never so clearly defined that intense sales competition was not taking place at their boundaries. Some firms might be increasing their sales, others just holding their own, while still others would be losing sales. These unequal rates of growth and decline might be related to internal structure and management, to

---

1/ This section draws heavily on Edmond S. Harris, 1966 and 1967, for the period up through the mid-sixties. See also Hedlund, 1964.

the impact of technological change, or to outside influences on the relative fortunes of the firms. In any case, they provided the impetus for constantly revising the spheres of competitive influence, either gradually and relatively peacefully or by direct incursion by one dealer into the sales territory of others. If an incursion into another's territory occurred, retaliatory sales competition could be expected, with the possibility of a price war. It was also not uncommon for a dealer's gradual encroachment on the business of others to culminate in retaliatory competitive action designed to reverse the process. Under normal conditions, competition among dealers not only skirted special influence areas but also avoided taking certain forms, most notably reducing the quoted price. The adherence by all dealers to a quoted price or price structure was, in part, a recognition of their mutual interest as sellers in relation to buyers. Secret price cutting, special rebates, discounts, or special services might seem to make the quoted prices more the exception than the rule. Nevertheless, the adherence by sellers to a uniform system of quoted prices was a matter of considerable practical importance to them. The quoted prices were the common guide for all of them and even the practices modifying the effect of uniform prices followed a pattern of restraint. The effect was to give each dealer the fullest advantage in negotiating with buyers who were, because of geographic location or lack of knowledge, to some degree isolated from other sellers. When dealers had to compete with each other, discounts and related practices provided a kind of price flexibility without jeopordizing each dealer's advantage with respect to his customers. Knowledge of accepted ranges of discounts and rebates, of the kinds of special services, and of the situations in which they would be offered became better known among the few sellers than among the many buyers. This gave a bargaining advantage to the seller when he negotiated to get or retain a customer.

The adherence by milk dealers to mutually acceptable quoted prices, about which they nevertheless engaged in a limited sort of price competition by means of discounts and services, reflected their general ambivalence toward competition and cooperation. The facts of business life required some forms of cooperation to avoid mutual destruction and to promote common interests. These forms of cooperation might be imposed upon some firms by the greater economic power exercised by other firms in the market. In any case, the result was a certain degree of order in the competition among the firms, which partially replaced the impersonal order imposed on sellers by a purely competitive market. These same mechanisms for cooperation could also provide the means for oligopolistic exploitation of the market by the imposition of excessive prices or distorted price relationships.

Several common forms of non-price cooperation among milk dealers included: (1) Establishment of a bottle exchange to facilitate the return of bottles; (2) arrival at a common understanding of trade practices conducive to the welfare of competitors; (3) promotional activities to advance sales and to stimulate favorable public opinion regarding the industry; and (4) negotiations with labor unions, producer organizations, and public regulatory agencies. In addition to these forms of cooperation, which could involve all dealers in a market, more restricted cooperative actions might involve a special group of dealers or just two dealers with a common problem. Thus, it was not unusual for milk dealers to help each other in relieving shortages of supply, or for smaller firms to agree on joint use of facilities (e.g., packaging equipment) that it would be uneconomical for one to use alone.

The most significant form of cooperation among milk dealers, however, was that which resulted in a recognized structure of prices and some tacit understanding of the manner and extent to which discounts and special services would be applied in competing for customers. This form of cooperation was influenced by laws and public policies that placed obstacles in the way of open and formal agreements. Written agreements were taboo. The process of arriving at a price structure or of amending prices became an interfirm activity that the participants tried to carry out with the same degree of privacy as activities within the firm. Meetings, luncheons, and telephone communications provided fairly effective means for firms to arrive at common understandings or to clear up misunderstandings on prices, customer relations, and other matters of common interest.

A certain "balance of competition" normally prevailed, rather tenuous in nature, maintained by the forms of cooperation described and by mutual respect for each dealer's sphere of special influence. The balance of competition did not prevent struggles for survival and growth from going on. Competition within the permissible limits might be intense and sporadically bitter. But as long as the competitive balance prevailed, prices were generally quite stable, and price changes were made in an orderly, concerted manner.

This balance of competition could be upset by changes that the market had no gradual and orderly way of assimilating. This might set in motion a whole series of changes, sometimes of cumulative intensity, and be accompanied by a marked disequilibrium of prices. When a balance of competition was again restored, marked changes often had taken place in the market. Some firms might be gone, the sales of those remaining reallocated, changes might have occurred in distributive techniques, and price relationships changed.

The changes in fluid milk processing and marketing discussed in the preceding sections—especially the increased mobility of packaged milk, the resulting larger packaged milk markets, and the increasing share of supermarket sales—made the competitive equilibrium in many city milk markets more subject to price warfare during the fifties and sixties. A price war occurs when two or more firms engage in retaliatory sales competition, in the course of which they cut their prices below the level of a previously existing price structure. Harris, in his intensive study of 27 price wars in 14 markets, classified them as follows (Harris, 1966, p. 72)—

o Price warfare as a phase of rough competition—1 case.
o Price warfare in relation to entry of a new firm—7 cases.
o Price warfare in relation to innovation—5 cases.
o Price warfare as a disciplinary device—2 cases.
o Price warfare as a device for reapportionment of the market—3 cases.
o Price warfare as an outgrowth of bargaining difficulties with the cooperative—1 case.
o Price warfare as a means of reorganization of prices and marketing practices—8 cases.

This classification is based on the primary causal factor involved in each instance. Other elements usually became involved as the price war developed.

Price wars had both destructive and constructive effects. The most important function was to provide a means whereby the existing price structure could be adjusted to changes in milk marketing. The usefulness of a price structure in promoting an orderly process has been discussed. It had to change as marketing conditions changed... if not gradually, then by an outbreak of price warfare.

Since the early sixties, the balance of power in establishing price structures for milk has shifted from dealers to retailers, primarily supermarket groups. The buying market facing fluid milk processors has changed drastically. The thousands of individual consumers on home delivery routes have been replaced by a relative handful of buyers for groups of stores, restaurants, and institutions. Increasing integration into fluid milk processing by major supermarket chains means that a significant portion of the market is foreclosed to other fluid milk processors—in 1980, 17 percent. Integrated dairy and convenience store groups foreclose another 6 percent. Central buying of fluid milk by retail groups who have not chosen to operate their own milk plants greatly reduces the number of buyers and changes the nature of the price-bargaining process. Bargaining now takes place more or less in private (except when the antitrust agencies take an interest) and only when the chainstore contract comes up for renewal.

Since a price war occurs when the firms in a market do not find it possible to adjust to changes in marketing conditions without recourse to what might be called commercial violence, the pattern of price wars over time provides an indication of the stresses on price structures (table 47). There were 15 price wars in 1954 and 1955 in these 81 markets. In the late fifties and the first half of the sixties, the rate increased, dropping off in the late sixties. There were an additional 19 price wars in 19 markets in 9 States which had resale price control for a part but not all of the period. The pattern of these is heavily conditioned by the timing of removal or imposition of resale price control, so they tell us something about the adjustment process only on a case-by-case basis.

Table 47--Number of price wars in 81 markets without resale price
control, 1954-69

1954-55	... 15
1956-57	... 12
1958-59	... 21
1960-61	... 25
1962-63	... 23
1964-65	... 22
1966-67	... 16
1968-69	... 9

Price wars in 81 markets without resale price control occurred
more frequently in middle-sized markets than in either large or small
markets. Probably the relative scarcity of price wars in large markets was due in part to timing...some of the adjustments in price
structures to increasing store sales and larger containers had already
been made by 1954. Perhaps the larger number of firms in bigger markets facilitated the adjustment of price structures without recourse
to price warfare. Reasons for the lower frequency in small markets
are more speculative.

Market size 1/	Number of price wars per market 2/
1-49	1.4
50-99	2.0
100-199	2.6
200-399	1.1
400 or more	0.7

1/ Milk sales (million pounds per month) in standard metropolitan
statistical area or county.
2/ Average number of price wars per market during the period,
1954-1969.

Price wars were a symptom of the adjustment process taking
place. In four States, resale price controls were instituted in
response to these strains. In five others, resale price controls
were dropped. In addition, a number of States instituted trade
practice regulation in an attempt to restrain the adjustment process.
These regulatory developments will be discussed in Chapter 10.

Price Structures

In markets for fluid milk products as they evolved, the results
of the price-making process are more aptly considered in terms of
structures than of prices. There is no one price in a market. At
the retail selling level, prices vary not only by type and size of
container, but often also between store and processor labels; sometimes between primary and secondary brands under store or processor
label; between supermarkets, convenience stores, dairy stores, and
small grocery stores; and often within these groups. There may be

as many as 25 different retail store selling prices of whole milk in a market or as few as four. A small number of these price differences are related to actual variations in physical characteristics of fluid milk--butterfat content, grade, vitamins added, etc. Some are cost-related, but more commonly, they reflect different merchandising policies of retailers.

Although price structures tend to be reasonably stable for extended time periods, one or more retailers--having a specific objective--will introduce a change--for example, a private-label differential or a secondary brand. Introduction of a new container such as a plastic gallon often triggers a readjustment in price structures. Discount food stores frequently attempt to use milk as a loss leader or at least a zero-margin item, upsetting existing price structures. Often this occurs when a new discount food store begins operations.

Dairy stores were once innovators in attempting to bring about drastic changes in price structure. In most markets, a modus vivendi has now been achieved. Frequently, supermarkets and dairy stores sell the basic grade of whole milk at the same price. In some markets, differentials existed for quite some time, often associated with the type of container--glass in dairy stores and paper in supermarkets. Most dairy stores, however, found that consumer bias against returning containers is more important than the cost advantages of glass in a captive operation, and switched from glass to paper and then to plastic.

Store price structures for whole milk vary widely between markets. In 1969-71, the number of prices for whole milk in stores ranged from 25 in Huntington, W. Va., and Washington, D.C., to 3 in Montgomery, Ala. (table 48). Markets under State resale price control, as was Montgomery, showed considerably less diversity in prices than did uncontrolled markets. Over 85 percent of the markets under resale price control had 10 or fewer prices, compared to 45 percent of the uncontrolled markets.

A base price can be identified in each market. In 1969-71, nearly always it was the supermarket price for half-gallon paper containers of basic grade milk--usually store-brand milk. Since then, plastic gallons have taken over the role. However, the proportion of milk sold at this base price varied greatly (table 49). In 16 of 56 markets, about a fourth of the milk was sold at the modal (most common) price. In another 15 markets, 30-39 percent of milk was sold at the modal price. In only one market was as much as 90 percent of milk sold at the modal price. In every market, at least some milk was sold above the modal price.

A considerable part of the variation in prices was, of course, due to differences among sizes and types of containers. Quarts universally sold at prices for two quarts which were 1-20 cents above half gallons in paper containers. Half-gallon glass generally cost less than paper, but not in all areas. Gallons generally were less than two half-gallon paper, with differentials ranging as high as 17 cents per half gallon-equivalent.

Pricing practices for store brands, compared with processor brands, varied widely. In most supermarkets carrying both store and processor brands, there was a differential between the two brands of

Table 48--Distribution of markets with specified numbers of prices for whole milk, by type of market, 56 markets, 1969-71

Number of prices	:	Controlled markets	:	Uncontrolled markets
	:		Number	
	:			
3	:	1		--
4-5	:	3		2
6-7	:	9		3
8-9	:	6		7
10-11	:	1		4
12-13	:	2		6
14-15	:	1		4
16-17	:	--		4
24-25	:	--		3
Total	:	23		33
	:			

-- means zero.
Source: Jones, 1973.

Table 49--Whole milk sold at the modal price, by type of market, 56 markets, 1969-71

Percent of milk sold in each market at modal price	:	Controlled markets	:	Uncontrolled markets
	:		:	
	:		Number	
	:			
0-19	:	0		1
20-29	:	3		12
30-39	:	5		10
40-49	:	3		3
50-59	:	5		4
60-69	:	1		1
70-79	:	1		1
80-89	:	4		1
90-100	:	1		0
Total	:	23		33
	:			

Source: Jones, 1973, p. 13.

milk in the same size and type of container, except in markets with State resale price control. In controlled markets, processor-brand and store-brand commonly sold at the same price. In uncontrolled markets, the most common differential between the two types of brands was 2 cents per half gallon, although they ranged up to 12 cents in Charleston, W. Va.

The most common practice of supermarkets that did not carry store-brand milk was to sell processor-brand milk at the price charged by other stores for store-brand milk. In a few markets, these supermarkets sold their milk at a differential above store-brand milk prices in other stores.

Dairy stores generally sold milk at prices below those in supermarkets, but not in every area. Dairy store prices for half gallons were below those in supermarkets in 9 of 11 controlled markets and 14 of 20 uncontrolled markets. The differential ranged as high as 13 cents per half gallon in New Haven, Conn., although the median differential was between 1 and 3 cents in uncontrolled markets and less than 1 cent in controlled markets.

There is considerable variation in the way that lowfat and skim milk products are priced in stores. Generally, these products sell at somewhat lower prices than whole milk. However, 2-percent milk sold at prices ranging from 1 to 10 cents more than whole milk in 20 percent of the uncontrolled markets and 1-percent milk was above whole milk in 42 percent of the uncontrolled markets selling the product. The most common differentials between 2-percent and whole milk were 2 cents and 5 cents per half gallon in uncontrolled markets and 1 cent and 4 cents per half gallon in controlled markets. Skim milk tended to sell at somewhat lower prices than 2-percent milk.

Price structures on home-delivery routes were even more diverse than those in stores. In part, this arose because of the variety in containers used on home-delivery routes. No single container was used in all markets. Half-gallon paper containers were most often used, closely followed by quart paper containers. Plastic was used much more in controlled than in uncontrolled markets.

Wholesale Pricing

The pricing process for wholesale sales of fluid milk products to hotels, restaurants, and other eating places, and for sale to small retail stores is generally of the quoted-price type. Occasionally, there may be some minor element of bargaining--particularly through the use of a threat to shift to another supplier. In general, however, such buyers are price-takers. Competitive moves are often nonprice.

The pricing process of chainstores and other groups differs. In general, they negotiate with processors to obtain a contract (either written or oral) to supply fluid milk products to a chain division or a substantial part of it. In general, it is fairly safe to assume that such prices are driven down to levels approximating the costs of providing the service. In many cases, they may not cover overhead. Activities of FTC in enforcement of price discrimination legislation tend to raise these prices above the level which would otherwise exist.

MARKETING MANUFACTURED PRODUCTS

Ice Cream

The growth of supermarkets markedly affected the merchandising of ice cream. In the thirties, drug stores sold most of the ice cream. After the introduction of the supermarket and prepackaged ice cream in half gallon containers, retail sales of ice cream rapidly shifted to supermarkets. In the sixties, specialty ice cream stores—most of them chain or franchised operations—entered the market in force, most of them selling relatively high-priced ice cream to consumers who preferred higher butterfat content or a different texture than that commonly sold in supermarkets. The specialty stores also featured a wider variety of flavors than found in most supermarkets. Subdealers are increasingly important in ice cream distribution, although not to the extent found in fluid milk. Large ice cream manufacturers, like fluid milk processors, are consolidating their manufacturing operations in fewer and fewer locations and establishing distribution depots from which deliveries are made. Often the closed ice cream plants become distribution depots.

The expansion of the market for soft-serve frozen desserts, primarily ice milk, through chain and franchised outlets such as Dairy Queen was rapid in the fifties and early sixties. The growth in the sales of frozen desserts was mostly in ice milk during this period.

Cheese

The retail cheese market today is drastically different from that of the thirties. Development of new types of cheeses and new processing and packaging methods led to the present supermarket cheese departments with several hundred varieties, types, and packages, in place of the handful of varieties from which the grocer cut a chunk in the old days. Natural cheese from many countries and seemingly endless varieties of processed cheeses, cheese food, and cheese spreads are available—all prepackaged and many of them sliced.

In the late sixties and seventies, chain and franchised wine and cheese shops featuring even a wider variety than supermarket cheese departments had quite a growth. Cheese appeared in many other specialty shops.

The marketing of processed cheese and related products is still largely in the hands of Kraft and Borden, although many others have relatively small shares of the market. A significant portion of the sales of Cheddar cheese is still made in random-cut chunks packaged at the chainstore warehouse. Other than this, most cheese is sold under manufacturers' brands.

There has been a dramatic growth in the market for Italian cheese primarily for use in pasta products. Here, most buyers are food manufacturers or operators of pizza places and Italian restaurants. Most pizza houses are chain or franchised operations.

The intermediate distribution system for natural cheese starts with assembly—collecting small diverse lots of cheese from individual manufacturers, sorting them, and aggregating into larger, more uniform lots. Grading and storage may be included. All firms performing these functions are called "assemblers." 1/

Specialized independent assemblers are down to about half a dozen in Wisconsin and none elsewhere. Most of their sales are to processors and intermediate handlers who cut and package natural cheese for distribution to wholesalers and retailers.

The large cheese processors, who are also the most important intermediate handlers, orginally bought cheese on the open market but they now assemble most of their cheese through a variety of arrangements. Many obtain a large proportion of their needs from their own plants or under informal sales agreements with proprietary and cooperative manufacturers. These processors are the major market for most regional cooperatives.

Agreements between processors and natural cheese manufacturing plants take many forms. The processors may take the entire output of selected manufacturers with a sufficient quantity of cheese. They take a specified amount from some plants on a regular basis and, when supplies are tight, they may make arrangements with additional plants for specific quantities. When cheese is abundant, assemblers may cut back on the amount of cheese they are willing to accept, in which case additional milk may be shifted into butter and powder.

The only cheese bought under an open offer under the price support program is in 40-pound blocks. However, the Government does provide an outlet for barrel cheese (in the form of process cheese, which is bought on a bid basis) but no open-end commitment exists to buy process cheese. In recent years, a number of plants have switched to the production of barrel cheese and cannot easily switch back to 40-pound blocks. Barrel cheese makes up a major portion of total Cheddar cheese production, since it provides economies in manufacture and package costs. This weaker guaranteed market position sometimes places cheese manufacturers in a weaker bargaining position and may explain the unwillingness of assemblers to accumulate unlimited quantities of barrel cheese.

Some types of cheese must be aged. Medium Cheddar cheese is usually aged 3-6 months and aged Cheddar longer. In the Italian-type varieties, Mozzarella is not aged, but Parmesan must be aged at least 10 months, Romano 5 months, and Provolone 60 days. Swiss cheese must be aged at least 60 days. Since aging requires large investments in storage capacity, most cheese is not stored at the manufacturing plant.

Some cutting and packaging is done at all levels, including manufacturing plant, intermediate distributor, and chainstore warehouse. National cheese companies and other intermediate handlers cut and package the greatest volume of natural cheese. Two of the major national cheese companies package primarily under their own labels, whereas the other national companies also package private labels for retail chains. Cheese cutting at the retail level used to predominate

---

1/ The remainder of this section is adapted from Lough, 1975, pp. 31-42.

but has declined with the introduction of rindless cheese and changing merchandising methods. Few retail chains cut and package cheese at the warehouse, although the number is increasing partly due to the growth of delicatessen departments in stores. Cutting and packaging gives the retail chain greater control over the type of product merchandised.

Over 80 cheese companies, including importers, have national distribution. About 40 of these are based in Wisconsin. Retail outlets take well over half of all cheese distributed. Institutional outlets are next, while industrial outlets have the smallest share. Both institutional and industrial markets are increasing their shares. The increase in institutional markets, which now use a large share of the process cheese, reflects the growth of fast food outlets. The industrial market uses mostly natural cheeses, except for filled cheese in frozen pizzas.

Butter

The principal changes in butter marketing include nearly universal packaging in one form or another and virtual disappearance of the old butter tub, the increased proportion of Grade A and AA butter (about 80 percent is now AA), more widespread use of consumer grades for butter as a merchandising tool, and the increase of private labels. As margarine has increasingly displaced butter, the butter market has become about evenly divided between the at-home and the away-from-home market. Restaurants which wish to convey a quality image use butter. Cafeterias, fast food outlets, and others with less concern for the quality image use margarine. Nearly all butter is now sold to these outlets as patties.

Specialized printing and packaging firms have been somewhat displaced by manufacturers who now perform these operations. Large plants, nearly all owned by cooperatives, now print and package most or all of their output. Smaller plants still sell some butter to independent packagers. (See Jones, 1977, pp. 19 and 22.)

Nonfat Dry Milk

Nonfat dry milk became a consumer product of some importance when the instantized product was introduced in the fifties. It is sold under private labels by many retailers and under a small number of packer labels of a few major companies. Growth of the home market ceased in the sixties and sales declined in the seventies.

Pricing Manufactured Dairy Products at Retail

Dairy products other than whole milk are not regarded by retailers as "competitive" products, that is, items on which they refuse to be undersold in price. Ice cream is widely regarded as an excellent traffic builder and is frequently specialed. It has been treated—like broilers—as a low-margin item for most of the post-World War II period. In fact, supermarket margins on standard ice cream became so low, due in part to frequent specialing, that intense efforts have been made more recently to build supermarket sales of "quality" ice cream at substantially higher prices and margins.

Cheese is frequently specialed. Specials on cheese are particularly useful since they attract buyers to the cheese display which may consist of a very large number of items, most of which carry a large margin. Thus, a special on a relatively inexpensive style of cheese may attract many impulse buyers to higher-margin items.

Butter is seldom specialed since it no longer possesses the transfer effect (of attracting buyers into the store who then buy other, "impulse" items) it once had. It is now a specialty item and the transfer effect of a special on butter is very small. Most butter buyers are dedicated to the product and will shift only with great reluctance.

Evaporated milk is specialed fairly frequently since it is often a large-budget item for consumers who use it in quantity. Thus, those using canned milk for baby formula can be attracted by a special when they may buy as much as a case.

Wholesale Pricing

Wholesale prices of processor-label dairy products other than fluid milk are set almost entirely by a quoted-price system. Larger buyers of private-label products can obtain products at negotiated prices, while smaller buyers deal with a quoted-price system.

Wholesale prices of butter and cheese fluctuate with the changing supply-and-demand situation, so that the pricing system for these products is something of a hybrid between the quoted-price system and supply-and-demand pricing. For most of the other products, prices fluctuate less often, being somewhat less sensitive to changes in supply and demand of raw milk. Butter is particularly sensitive to changes in supply and demand because of its residual nature. Wholesale prices of butter, nonfat dry milk, and American cheese rest on a floor provided by the support purchase program of the U.S. Department of Agriculture so long as the Department is purchasing these products. When supplies become tighter, prices tend to rise above support levels.

Butter. Prices received for butter by manufacturers, primary receivers, and others at the wholesale level are based primarily on activities on the Chicago and New York Mercantile Exchanges. Prices received by manufacturing plants for bulk butter are based almost exclusively on spot prices on these exchanges less freight charges to Chicago or New York. In addition, a plant may receive a premium for such factors as uniformity, size of shipment, or a special flavor characteristic. Generally manufacturers who sell only bulk butter are price-takers, not price-makers. 1/

Manufacturers who soft-print and package butter sell it to primary receivers, grocery chains, dairies, and restaurants. These

---

1/ This section on butter pricing is based on Jones, 1977, pp. 24-26; Hammond, 1967; and Natl. Comm. on Fd. Mktg., 1966.

manufacturers may, depending on competitive conditions, receive a better return than those who sell only bulk. They do not absorb the freight cost for printed butter as for bulk and they are in a position to bargain over the markup charge ("overage") for printing and packaging.

Primary receivers buy butter from manufacturers at spot market prices (plus possible premium) and sell to several types of customers. Print butter is sold to grocery chains and wholesalers who supply retail food stores. Bulk butter is sold to other primary receivers, butter wholesalers, food processors, and cold storage firms. These sales are based on spot prices plus markup to cover handling, overhead, and profit.

Primary receivers are both price-takers and price-makers. Prices they receive (and pay) for butter are based on spot market prices, but since many of the receivers are members of the Exchanges, by buying and selling butter on these exchanges they can influence the spot market price. Nonmembers can also buy and sell on the exchanges through brokers.

Cheese. Pricing methods for cheese leaving the manufacturing plant in wholesale lots vary with marketing practices. 1/ Sales transactions at the cheese plant are either spot sales or contract sales. Spot sales include a very small part of the total cheese volume traded and are reflected in National Cheese Exchange and assembly point prices. They involve transactions between large first handlers of cheese or sales from manufacturing plants not contracted to a first handler. Long-term contract sales between a cheese plant and first handler represent the bulk of cheese trading at this level.

The National Cheese Exchange in Green Bay, Wis., meets every Friday and provides a place for members and owners of licensed cheese factories to buy or sell 40-pound block Cheddar, barrel American, and Swiss block cheese. In 1978, the Exchange had about 44 members who handled 80-90 percent of all the cheese marketed in the United States. The volume sold on the Exchange represents only a small proportion of those members' supplies and has never exceeded 1 percent of U.S. production.

The Exchange serves as a source of extra supply and a place to dispose of surplus. It is an alternative market, since most trading in cheese is done through regular commercial channels. Its primary function is to provide a national index of the value of cheese, reflecting supply and demand conditions. The prices arrived at on the Exchange on Friday are the basis for nearly all transactions in American and Swiss cheeses, and some Italian, blue, and other specialty cheeses, for the following week.

Assembly point prices represent spot trading at points where the first handler may obtain alternative supplies resulting from a short-run supply and demand imbalance. Premiums are negotiated for each

---

1/ This discussion of cheese pricing is based on Lough, 1980.

shipment separately and fluctuate more than premiums under a long-term contract. Changes in these premiums are often the first indication of a changing cheese market (Leathers, 1982). Thus, small price changes at Wisconsin assembly points may give an early indication of price movement. In a tight supply and demand situation the mid-point of the range in assembly point prices may be several cents per pound over the Exchange price plus the assembling charge, whereas in a "flat" market there may be no difference. This may increase the Exchange price some time in the following weeks, after which the assembling point price will further increase.

Long-term contracts between cheese plants and national cheese marketing firms are common. In these, all or a specific proportion of a plant's production is included, with the price based on the Exchange price plus a specified premium.

Retail stores are primarily serviced by brokers who take orders in an area, sometimes in combination with sales representatives of the manufacturers. A few large companies employ representatives who visit individual stores and some companies employ brokers who have warehouses and trucks and service individual stores in a market area.

Most manufacturer's brand cheese is sold to retailers on the basis of a weekly price list for nonspecialty cheeses which follows price changes on the Exchange. Firms with major brand cheeses follow less closely, considering factors such as changes in other costs and in target margins. Some sales of private label cheese to large retail customers are also based on a price list (less advertising and promotion costs), but most are on a formula using the Exchange price plus a premium.

Sales to full-line institutional wholesalers commonly use weekly or monthly price lists. These are based on the Exchange price plus a premium, which may remain unchanged for up to a year if there is little change in marketing costs or the quantities of cheese required during the year. Sales to large-volume chain food service customers who require special cheeses often use a long-term formula price based on the average Exchange price for the preceding month plus a premium. Most first handler sales to industrial markets use a formula based on the Exchange price on the date of cheese production plus a premium (Hayenga, 1979).

CONSUMER MARKETS

Consumers--the customers who are the reason for the existence of the entire food system--are as different from those of the Depression years as are farmers. Galbraith's phrase "the affluent society" is the best short characterization of the present era. Affluent does not mean rich, at least by American standards, but it does mean well past the subsistence stage.

Total consumer income rose faster than food prices for most of the period since 1947. This means that the percent of total income spent for food declined or was unchanged in all but two years since then. Over the entire period, it decreased from 26 percent to 16 percent.

The general picture furnished by this old reliable statistic is correct. But the personal consumption expenditure series of the Department of Commerce which is the basis for these computations does not tell the entire story. 1/ Comparing total food expenditures with the cost of three different food plans developed by the Agricultural Research Service provides a comparison of the amounts that consumers and the entire society decide to spend for food with the costs of fixed quantities of food at three different levels. Each of the three food plans provides a nutritionally adequate diet. They are designed so that the low-cost food plan costs about 80 percent as much as the moderate-cost food plan and the liberal food plan about 20 percent more (Peterkin et al, 1975). These relationships vary a bit from year to year, but the general picture is accurate over time. Aggregating the cost of each food plan for the population of 1960 and that of 1976 gives the following picture:

	1960	1976
	Mil. dol.	
Aggregate expenditures for the U.S. population at levels of:		
Low-cost food plan	56,535	135,663
Moderate-cost food plan	71,553	170,281
Liberal food plan	86,208	204,624
Actual expenditures at retail store price levels	71,006	177,293

In 1960, actual expenditures for food (with all food at retail store prices, as it is in the ARS food plans) was very close to the aggregate cost of the moderate-cost food plan. By 1976, actual expenditures had risen well above the level of the moderate-cost food plan. They had moved 20 percent of the distance from the moderate-cost food plan to that of the liberal food plan.

Thus, we see that, while consumers on the average were well past the subsistence level in 1960, in the next 16 years they had moved significantly toward better eating, i.e., eating more expensive foods. In addition to the shift shown in these figures, they were also eating a significantly larger share of their meals away-from home at substantially higher cost.

Total quantities of food per person do not vary greatly from year to year or over the long run. The old saw about the limited size of the human stomach does have some applicability. What does change is the composition of the diet. The proportion of meat and other higher-priced foods is significantly greater than it was in earlier times.

Between 1954 and 1976, food expenditures per person increased 150 percent. Most of the increase was due to rising prices, but shifts in outlets--mainly to away from-home eating--and shifts among foods were also significant.

1/ For a more comprehensive discussion of the Commerce series, see Manchester and King, 1979.

116

Precent of increase due to--

Changes in prices 73
Shift in outlets 11
Shifts among foods 10
Changes in quantities 6

The share of food dollars going for meals and snacks away-from-home more than doubled, although the increase is not nearly as great when discounted for differences in price levels and price movements (table 50). In the food-at-home market, leaving aside home production and donated foods, the supermarkets' share nearly doubled between 1958 and 1980 (table 51).

Convenience stores increased their share of the market from almost nothing in 1958 to 4 percent in 1980. The share of smaller grocery stores declined fairly steadily for two decades.

The rise of the supermarket and the decline of smaller grocery stores reflect many factors including the demise of the independent grocery store and the growth of chains, cooperative and voluntary groups, and franchise operations. Other food stores--meat markets, fruit and vegetable stores, bakeries, dairy products stores--declined until the mid-sixties but increased modestly afterward.

Table 50--Expenditures for food at home and away from home

Year	Sales	Home production and donations	Total	Sales	Supplied	Total	All food
		Food at home			Food away from home		
			Percent				
1929	68.0	13.9	81.9	15.5	2.6	18.1	100.0
1939	66.9	13.0	79.9	16.8	3.3	20.1	100.0
1948	65.2	10.2	75.4	20.2	4.4	24.6	100.0
1955	67.8	6.5	74.3	21.2	3.7	25.7	100.0
1960	68.6	5.2	73.7	22.2	4.1	26.3	100.0
1965	66.2	3.8	70.0	25.9	4.1	30.0	100.0
1970	63.4	3.1	66.5	29.1	4.4	33.5	100.0
1975	61.7	2.8	64.5	30.7	4.8	35.5	100.0
1980	59.9	2.3	62.2	32.9	4.9	37.8	100.0

Source: Manchester and King, 1979, and revisions.

Table 51--Sales of food for home use

Type of seller	1939	1958	1977
		Percent	
Supermarkets	5.8	36.5	63.2
Convenience stores	0	.1	3.2
Other grocery stores	52.4	37.0	16.9
Other foodstores	14.1	11.2	7.7
Other stores	7.7	4.7	5.9
Home delivered	14.9	7.3	1.3
Farmers, processors, wholesalers, other	5.1	3.2	1.8

In the away-from-home market, the big increase was in sales, as distinguished from food furnished. The big gainers in market share in recent years have been fast food places. Their share rose from 5 to 29 percent in 19 years (table 52). The share of schools and colleges rose in the sixties and then declined as school enrollments leveled off. While dollar sales increased for all types of outlets, the shares of other sellers declined or grew slowly because of the rapid rise in sales by fast food places.

Not only did retail outlets change, but also the types of organizations within them. Supermarkets started by independents in the thirties to combat the grocery chains are now mostly operated by chains or by members of wholesaler-sponsored voluntary groups or retailer cooperatives. Convenience stores are almost entirely chain or franchised operations, as are fast food outlets.

Table 52--Shares of away-from-home food market sales

Type of seller	1939	1958	1977
		Percent	
Restaurants, lunchrooms, cafeterias	46.6	53.5	38.5
Fast food places	7.1	5.4	29.3
Hotels and motels	10.8	6.1	5.4
Schools and colleges 1/	6.8	12.0	11.1
Stores, bars, direct selling	21.1	14.7	9.2
Recreational places	1.9	2.4	2.3
Others	5.7	5.9	4.2

1/ Includes child nutrition subsidies.

All of these changes had their impacts on the markets for dairy products. Consumers changed the foods they bought and the places where they bought them because, as a group, they had the means--both economic and technological--to do so.

Consumption of Dairy Products

The fifty-five years of dairy products consumption shown in table 53 cover the affluence of the twenties, Depression, immediate postwar, and much more. At least four major forces were at work:

        o Prosperity and depression.
        o Baby boom(s).
        o Commercialization of dairy farming.
        o Cheaper substitute products.

Canned milk consumption went up during the Depression as a lower-priced substitute for fluid milk and stayed up into the fifties in babies' bottles. Rising incomes, the end of the baby boom, and substitution of infant formulas and mother's milk combined to lower consumption. Packaged nonfat dry mik enjoyed a minor boom with the introduction of the instantized product but an affluent society has turned away from it partly because it is no longer the bargain it used to be.

Butter sales declined throughout the period, largely as a result of competition from lower-priced margarine. The rate of decline has slowed in recent years.

Cheese sales rose in every decade, soaring between 1966 and 1976. The biggest increases were in pizza cheese, reflecting the large group of teenagers and young adults in the population, the boom in fast food eating places, increased incomes, and a shift in demand.

Beverage milk and frozen dessert sales reflect the age structure of the population with children drinking more milk and eating more ice cream than adults. As the population grew older after the mid-sixties, milk sales have declined and frozen dessert sales somewhat less.

Fluid cream lost markets to substitutes throughout the postwar period. Increasing sales of yogurt and sour cream have pulled up the group totals in the last fifteen years.

Since 1946, per capita consumption of dairy products has dropped 30 percent, while population increased 64 percent. Total consumption rose 15 percent. But only 30 percent of the decline in per capita consumption was due to a drop in sales to consumers, restaurants, and institutions. The rest was due largely to the near-disappearance of the farm with one or two cows producing milk for home consumption. There are fewer farmers now and not many keep a cow or two to produce milk for use of their own families. So the contribution of that milk consumed where produced to total dairy product consumption is only 4 percent of what it was in 1946.

At the same time, we have had increased efficiency of utilization of the milk supply, primarily of solids-not-fat. In the mid-thirties, much of the milk was sold as farm-separated cream and half

Table 53--Per capita use of dairy products 1/

Product	1926	1936	1946	1956	1966	1976	1981
				Product pounds			
Sales of consumer products:							
Beverage milk	151.3	169.7	266.9	268.4	268.0	249.5	233.0
Cream and specialties:	7.1	8.0	10.0	8.1	6.8	7.7	8.4
Cheese	4.6	5.4	6.7	7.4	9.4	15.4	17.4
Cottage cheese	.6	.9	2.1	4.3	4.5	4.7	4.3
Frozen desserts	9.8	9.9	24.7	24.1	28.2	27.6	26.2
Butter	14.4	13.6	8.6	7.4	5.4	4.3	3.9
Canned milk	10.7	14.5	17.6	14.3	8.2	3.6	2.9
Packaged nonfat dry milk	0	0	0	.8	1.3	1.0	.6
Total	198.5	222.0	333.7	334.9	332.0	313.8	296.7
Sales of ingredient products:							
Bulk con- densed milk:	3.5	4.1	10.5	6.2	7.1	4.8	4.2
Dried and malted milk: and whey	.8	2.1	4.5	4.5	5.1	5.6	5.5
Total sales:	202.7	228.2	351.7	345.6	344.2	324.1	306.4
Consumed where produced	137.0	115.9	92.2	47.0	16.1	6.4	3.9
USDA donations:	0	0	0	2.0	1.2	.3	1.3
All dairy products	339.7	344.2	443.9	394.6	361.5	330.9	311.6

Detail may not add to totals because of rounding.

1/ Contribution to average consumption per person resident in the U.S.

of the total solids-not-fat was fed to animals or wasted. The figure is now down to 17 percent. This was due in large part to the growth of the nonfat dry milk industry in World War II continuing into the fifties and, more recently, to increased use of whey. Thus, any given level of milk production now yields substantially more human food than it used to.

The dairy problem is sometimes described as a problem of under-consumption rather than overproduction. But these figures show that 71 percent of the drop in per capita consumption between 1946 and 1981 was due to a decline in use on farms where produced and only 29 percent to a decline in sales. The big sales decline was in butter, because of the competition from margarine. This decline has nearly bottomed out. There was an even larger drop in evaporated and con-densed products. These are now mostly ingredients in other food prod-ucts (including dairy products) and they have been under strong competitive pressure from whey, soy products, and caseinates.

Most of the impression of a dramatic decline in demand for dairy products results from looking at consumption rather than sales. Sales of total solids in all dairy products--excluding milk produced and consumed on farms, donations from CCC stocks, and subsidized consump-tion in the National School Lunch and Special Milk Programs--peaked in the late fifties at 60.5 pounds per person, declined during the sixties, and leveled out in the seventies (table 54). Butterfat consumption declined in each period until the last few years, while solids not-fat rose until 1960-64 and then declined slowly.

Table 54--Commercial sales per capita of butterfat and solids-not-fat in all dairy products 1/

Period	Butterfat	Solids-not-fat	Total solids
	Pounds		
1947-49	25.2	33.2	58.4
1950-54	23.9	35.7	59.6
1955-59	23.0	37.5	60.5
1960-64	21.1	38.1	59.2
1965-69	19.9	36.3	56.2
1970-74	18.9	36.6	55.5
1975-79	18.9	35.6	54.5
1980-81	18.6	34.2	52.8

1/ Excludes milk produced and consumed on farms, donations from CCC stocks, and subsidized consumption in the National School Lunch and Special Milk Programs.

IN SUMMARY

Marketing of packaged fluid milk shifted from three-fourths home delivery in 1929 to less than 7 percent today, as the entire food retailing picture changed with the growth of supermarkets. Milk containers changed from mostly glass quarts to a high proportion of plastic gallons. New products included homogenized milk, half and half, lowfat milk, yogurt, and sour cream. Supermarket distribution changed to private label milk bought under contract in the sixties and seventies. Dairy and convenience stores were started in most markets.

The nature of competition changed sharply as these developments occurred. Dealers had been generally dominant up into the sixties but power shifted to supermarket groups by the seventies.

The marketing systems for manufactured dairy products also changed. Ice cream became a supermarket special item, where once it was mostly sold in drug stores. Cheese varieties and brands proliferated, although the two large cheese companies continued to play a central role. Intermediate distribution of cheese and butter became more integrated. Wholesale pricing systems for butter and cheese continued to be based on exchange trading.

Per capita consumption of all dairy products has been declining since the twenties, but most of the decline is in farm home consumption of milk and butter. Substitutes have taken their toll, most notably margarine. The postwar baby boom created a bulge of young people in the population and a similar one in milk consumption. In recent years, total commercial use of milk in all products (but not including CCC-donated foods) has been reasonably steady, with population growth about offsetting declines in per capita use.

Part 3

Institutions
Affecting Price Formation

8
The Concept
of Orderly Marketing

The Agricultural Marketing Agreement Act of 1937 makes "orderly marketing" one of the objectives to be served by marketing orders and agreements. The term has caused many problems to commentators over the years. It has been discussed by a variety of persons (see Nourse et al, 1962, for one). It is sometimes dismissed as a bit of legislative rhetoric with little meaning in the past and none now. The term came from outside milk marketing but the concept was and still is important in milk marketing and pricing, being closely related to the inherent instability discussed in Chapter 2.

DEVELOPMENT OF THE IDEA

In recent years, "orderly marketing" has come to mean, in many agricultural circles, the programs carried out under Federal marketing orders. This is a severe limitation, if not a perversion, of a much broader concept. 1/

The term "orderly marketing" began to acquire a specific meaning in connection with farming shortly after the end of World War I. Very simply, it was to spread the marketing of a farm commodity over a sufficient period of time to avoid the local drops in prices often occurring when large quantities of a commodity were thrown on the market at harvest time.

Cooperative marketing seemed to offer possibilities for orderly marketing. Some fruit and other specialty crop growers had organized successful cooperatives in California in the early 1900s. These had been successful, at least in part, in maintaining an even flow of their crops to the market.

The idea was developed by a young California lawyer, Aaron Sapiro, who concluded by 1920 that what would work for California's specialty crops could be almost universally applied.

Sapiro spoke to the recently formed American Cotton Association on April 21, 1920, and presented an eight-point plan for commodity cooperative marketing. His seventh point read: "Orderly marketing

---

1/ Based on materials by Wayne D. Rasmussen, ERS.

of the product throughout the whole production period." Later in the year, Sapiro carried his message to the wheat growers of the Great Plains and the Burley tobacco growers of the Southeast. By the end of the year, national associations of cotton, wheat, and tobacco growers had been formed to carry out the orderly marketing of their crops.

The term "orderly marketing" became the particular watchword of Sapiro's addresses. And, as often happens, he gave it particular meaning. It included establishment of market standards and grades, reduction of inefficiency in distribution, searching for more adequate utilization of commodities, and aggressive marketing.

During the twenties, the movement toward orderly marketing through cooperatives was seen as an alternative to the McNary-Haugen bills with their two-price system. When the Curtis-Crisp Bill was introduced in Congress in January 1927, Secretary of Agriculture William Jardine said that its aim was to promote orderly marketing, particularly of non-perishable products.

The Agricultural Marketing Act, an outgrowth of the Curtis-Crisp bill and similar proposals, was signed on June 15, 1929. One of its purposes was to prevent and control surpluses "in any agricultural commodity, through orderly marketing and distribution." The Federal Farm Board, established by the Act, aided cooperatives for many commodities, but the Great Depression and the lack of authority to control production doomed it to failure.

Neither the Agricultural Adjustment Act of 1933 nor its amendments of 1935 contained the term "orderly marketing." The 1933 act gave the Secretary of Agriculture authority "To enter into marketing agreements with processors, associations of producers, and others engaged in the handling, in the current of interstate or foreign commerce of any agricultural commodity or product thereof."

In 1937, Congress reenacted the marketing orders and agreements section of the 1935 amendments as the Agricultural Marketing Agreement Act of 1937. In its declaration of policy, Congress returned to the concept of "orderly marketing" by stating that "the disruption of the orderly exchange of commodities in interstate commerce impairs the purchasing power of farmers..." It called for reestablishing "orderly marketing conditions for agricultural commodities in interstate commerce as will establish, as the prices to farmers, parity prices..."

In 1954, Congress went back to the original concept by declaring its objective, in an amendment to the 1937 act: "Through the exercise of the power conferred upon the Secretary of Agriculture under this title, to establish and maintain such orderly marketing conditions for any agricultural commodity enumerated in Section 8c(2) as will provide, in the interests of producers and consumers, an orderly flow of the supply thereof to market throughout its normal marketing season to avoid unreasonable fluctuations in supplies and prices."

ORDERLY MARKETING OF MILK

The Federal market order system was created during a period of extreme disorder at the bottom of the depression to help create orderly marketing conditions. If fluid milk markets were to have an orderly supply, there had to be orderly production. For orderly production--both efficient and remunerative--there had to be: (a) provision for orderly physical assembly and distribution; (b) dependable and equitable contract relations between handlers and producer organizations and between these organizations and their individual members; and (c) orderly relationships as to prices and supplies between different markets. 1/

The structures and practices devised by cooperatives and handlers and the complementary mechanism of regulation supplied by government, taken together, have gone a considerable distance toward achieving these ends. The basic concept of "orderliness" in the economic sense includes: (a) Order in the time dimension--as related to seasons, longer cycles, and long-run changes in productive conditions and consumptive demand; (b) Order in the geographic dimension--within a market area, within a region embracing several such areas, and in national areas of industrywide adjustment--including economic adjustment between the two branches of dairy farming, fluid milk and milk for manufacturing uses; and (c) Order in competitive relationships--or equity--equal treatment of firms equally situated. Put another way, the last means no unfair price advantage to any handler or group of handlers.

Early cooperatives viewed orderly marketing as a local rather than a national objective. This view was pervasive from the beginning in the drafting and administration of milk marketing orders. But something less parochial emerged in the sixties--a recognition that the outlook of the Secretary of Agriculture and his aides should be industrywide and national in its scope. Hence, the Secretary is empowered to develop a system of fluid milk marketing orders, integrated as to their relations with each other and with all the uses into which milk goes, not merely orderly as to their internal housekeeping.

Most of the attention of analysts has been devoted to the time and space dimensions of orderliness. Indeed, Edwin Nourse did not find it necessary to even mention the competitive dimension. The requirement that similarily situated handlers pay at least the same prices for milk in the same use appeared then to have taken care of the problem.

Classified-use pricing was developed mostly in the twenties largely to deal with the flat-price problem. Classified pricing was utilized by cooperative milk bargaining associations in many larger markets in selling to dealers. However, the dealers who bought from independent producers generally continued to pay flat prices comparable to the blended returns received by the members of the cooperative associations (Spencer and Christensen, 1954, p. 36). Such dealers were often those who engaged only in the distribution of fluid milk or had a relatively small proportion of milk used for manufactured products. The flat-price buyer was buying milk for

---

1/ The first two paragraphs are adapted from Godwin et al, 1975, pp. 6-7, and Nourse et al, 1962, pp. 9-10.

fluid use cheaper than his competitors (if he did not incur higher
costs of balancing his own supply and demand than did his competitors, particularly where the cooperative was doing the balancing)
and thus could cut retail prices and increase his business. The
cooperative, in turn, in order for its buyers to compete effectively
with the cut-price milk, was also forced to reduce its Class 1 price.
But this reduction in Class 1 prices also reduced the blend price it
could pay its producers and allowed the flat-price buyer in turn to
further reduce his price to producers. Once the succession of price
cutting started, the price to producers usually declined to about the
level of prices for manufacturing milk. Producers of grade A milk
no longer were compensated for the extra cost of producing it (U.S.
Agric. Adj. Admin., 1939b, p. 25). In such a situation, imposition
of Governmental regulation--whether Federal or State--requiring that
all dealers pay the same price for milk in the same use eliminated
the advantages of a flat price buyer. This was essentially the
situation in 1961 when Nourse discussed orderly marketing.

Transfering the functions of bulk milk marketing from proprietary handlers to cooperatives in the sixties and seventies largely
shifted the cost of operating the bulk milk marketing system from
handlers to cooperatives and led to the imposition of over-order
charges by cooperatives (see Chapter 7). This development again
made it possible for one processor to buy milk for fluid use at a
different price than his competitor. The processor who did not buy
his milk from a cooperative, which by now had fallen heir to the
balancing function for the entire market, was able to avoid paying
his share of the cost of such a system. By buying directly from
independent producers or from a limited service cooperative, he
could purchase milk for use in fluid milk products at a lower price
(see Jacobsen, 1978, for an example).

GEOGRAPHIC PRICE STRUCTURE

The economic principles which constrain the policymaker in lay-
ing out a geographic structure of Class 1 prices are derived from
spatial economics. Strictly speaking, spatial economic theory is
developed in terms of long-run equilibrium conditions toward which a
free market is always being pushed, with frequent changes in the long-
run equilibrium as conditions change. But there is also a "correct"
solution at any point in time, given the existing quantities supplied
and demanded. This short-run spatial equilibrium is such that, if
free flow is permitted, no shipper can improve his returns by reallo-
cating his supplies among the outlets (i.e., the net revenue of each
shipper is maximized) and transfer costs are minimized. (See Bressler
and King, 1970, esp. p. 100, and Bressler, 1958, p. 116.) The basic
principles are as follows:

o In a free-flow market with spatially separated fluid milk
    markets in a "sea" of manufacturing grade milk, prices would
    rise above the manufacturing milk price surface which is
    relatively flat because of lower transport costs on manufac-
    tured products than on bulk milk (the dotted lines are con-
    tinuations of the manufacturing milk price surface). At the

outer boundary of the milkshed of each fluid market, the price for fluid grade milk would be higher by the cost of meeting fluid grade sanitary standards and the added costs of marketing bulk milk for fluid use. Prices would increase to the market center (the "peak" on each of the "houses") by the cost of transporting fluid milk (Figure 3a).

o If the "sea" of manufacturing grade milk did not cover the entire country or was not evenly distributed, there would be a price surface for manufacturing grade milk reflecting the spatial distribution of production and demand for products made from manufacturing milk (Figure 3b). The uneven distribution of production of manufacturing grade milk is the cause of the uneven price surface ilustrated.

o When fluid markets grow together, prices at the boundary are equal (Figure 3c). If one is much larger than the other, they merge and become one (Figure 3d). [The dotted lines are the market centers of the merged market (A) and the former secondary market (B).]

o The ceiling or limit price in any market is the delivered price (i.e., including transportation) at which milk for fluid use could be shipped from the nearest point having available supplies. This applies to all markets, but can be seen most clearly in the case of markets isolated by natural conditions—say a desert—from other areas where milk production takes place.

o The minimum price of milk for fluid use is set by the local supply function—if production costs are high because of local conditions (shipped-in feed is expensive, for example), prices will be well above an extension of the manufacturing milk price plane plus average costs of meeting sanitary standards. The limit on prices in the isolated market is the cost of fluid product milk from the nearest production area including transportation costs (if transportation is technically possible). (Figure 3e).

o A more rigorous statement of the economic model involved is given by Babb et al (1979, p. 18):

> "Such a model would equate supply and demand
> for all [markets], net of transportation
> costs. Prices between markets would differ
> by no more than transfer costs. These con-
> ditions are illustrated in the familiar
> back-to-back diagram of spatially separated
> markets where excess supply functions are
> derived."

o The supply function relates to the blend price (the average price of milk in all uses), so the Class 1 price for milk used in fluid products must be high enough to cover the reserves needed to meet short-run variations in supply and demand. Since the value of reserve milk used in manufactured

Figure 3a   Spatially Separated Fluid Milk Markets in a Sea of Manufacturing Milk

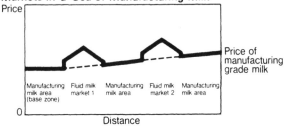

Figure 3b   Spatially Separated Markets of Uneven Size and Distribution

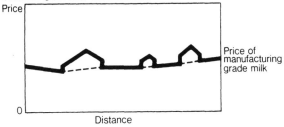

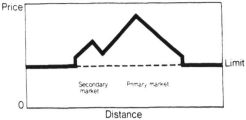

Figure 3c   Fluid Milk Markets Grow Together

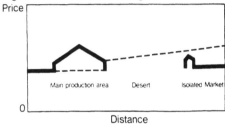

Figure 3d   Markets Merged

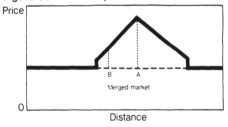

Figure 3e   Naturally Isolated Market

products is set by the manufacturing milk price surface, the differential between the prices for milk in fluid products and that for milk used in manufactured products will be large compared to that in the main production area.
o If market power is introduced--say by a cooperative with 100 percent membership--in the isolated market, any price up to the limit is possible.

The geographic structure of Class 1 prices which one would anticipate in a competitive market on the basis of economic location theory has these characteristics: From the major surplus production area (surplus with respect to fluid milk needs), prices would rise to more distant markets, reflecting transportation costs and local supplies and demands. The largest supply of milk over and above local needs is in an area extending from Minnesota to New York and Pennsylvania. Thus, any market in the United States could be supplied from some point in that zone at the base price plus transportation. So, as long as milk can move freely from one area to another, that price plus transportation would set the upper limit on prices in any market. If the supply in a market area elsewhere in the country is greater than local needs, the price at that point would be lower than base-zone-plus-transportation by an amount large enough to move the milk to a point where it is needed. With large surpluses on the West Coast, prices there would be no higher than in the eastern base zone. The level of prices on the West Coast would depend, of course, on the supply and demand functions in that area. It could be anywhere between the base zone price and the limit price. The existence of large surpluses (CCC purchases at the support level) would indicate that supply and demand conditions were such that the base price would be no higher than in the base zone. The California (State) pricing system now in use is based on a recognition of these facts and principles.

The principle of comparative advantage and the economics of location indicate that, in a competitive system responding to economic forces, milk for fluid use (including a reserve to meet day-to-day and seasonal fluctuations) would be produced near consumption centers, if it can be produced at or below the cost of milk from the base zone including transportation. Milk for use in manufactured products would generally be produced in the base zone. Thus, if the structure of prices brings forth large volumes of milk for use in manufactured products over the amounts needed for a fluid milk reserve in areas outside the base zone(s), the geographic structure of prices is not aligned in accord with economic principles. Base zones are not limited to one--they can exist anywhere in the country.

When supplies are tight, as in 1973, surpluses above reserve requirements are small everywhere except in the Upper Midwest, the Northeast, and on the West Coast. When production increases, supplies above fluid-needs-plus-reserve develop in many areas. This means that a geographic price structure which is appropriate (in the short-run, not in equilibrium) at one time is not appropriate under other supply-demand conditions. When supplies are very large, as they were in the mid-sixties and are now, a relatively flat price surface is appropriate--rather like a soup bowl which is flat on the bottom and only rises toward the edges. (See Hallberg et al, 1978.) When supplies are tight, the appropriate price surface at that time more nearly approaches the inverted cone shape of base-zone-plus-transportation.

IN SUMMARY

The concept of orderly marketing was developed in the twenties as a broad program to regularize the flow of agricultural products to market and increase the incomes of farmers by giving them market power through cooperatives. At the same time, milk marketing cooperatives were applying comparable ideas through classified-use plans for fluid milk markets. Order in milk markets came to have meaning in three dimensions--time, space, and competitive relationships.

9
Public Programs
During the Depression

Federal dairy programs were depression-born. The only previous attempts to affect the markets for milk and dairy products had been during World War I, primarily an effort to hold down prices, and some abortive efforts by the Federal Farm Board. During World War I, the United States Food Administration controlled milk, butter, cheese, condensed, evaporated and powdered milk. In the beginning, price fixing was avoided. Margins and profits were required to be reasonable and, in some cases, maximum margins were set. Regional Milk Commissions were appointed by the Food Administration for New York, Boston, Chicago, Philadelphia, and Ohio. The Regional Commissions had no clear-out statutory authority to fix prices. They attempted to reach voluntary agreements through mediation, without signal success. Rapidly changing costs and fluctuating demand because of the war frustrated the efforts. At least in New York and Boston, they ended up fixing prices for a time (Cadwallader, 1938, p. 799; U.S. Fed. Trade Comm., 1921, pp. 115-126; Pearson, 1919, p. 91; Spencer and Blanford, 1973, p. 201; Gilbert, 1919; Spencer, 1972; Guth, 1981).

Classified pricing plans operated by cooperatives were breaking down badly by early 1933. Cooperatives did not have the power to audit the records of processors who purchased milk from them to determine whether or not they were correctly reporting the uses of the milk. In other words, underpayment was widespread, as found in the Federal Trade Commission investigation (U.S. Fed. Trade Comm., 1937, pp. 74-75).

In addition, cooperatives seldom if ever had all of the milk in a market under their control. Even after the beginning of Federal regulation, cooperatives in half the markets supplied 70 percent or less of the milk (table 55). In none of the 52 markets for which licenses or orders were issued in the early years of the Agricultural Adjustment Administration were all distributors following the classified price plan (Gaumnitz and Reed, 1937, p. 167). Spencer estimated that 40 percent of the milk in the New York market was bought on a flat price prior to public regulation (Spencer, 1951, p. 22).

Table 55--Proportion of producer milk (excluding producer
distributors) purchased by distributors from cooperative
members, 26 Federally regulated markets, 1934-35

Percent of milk purchased from cooperative members	:	Number of markets
91-100 percent	:	5
81-90	:	9
71-80	:	1
61-70	:	5
51-60	:	4
31-40	:	1
0-10	:	1
Total	:	26

Source: Gaumnitz and Reed, 1937, p. 28.

SANITARY REGULATION

Sanitary regulation by the States and municipalities was well
established in most larger places by the thirties. The late-comers
to such regulation, especially in the South, tended to adopt the
USPHS standard milk ordinance.

There can be little doubt that local sanitary regulations and
State regulations, perhaps to a lesser extent, were widely used
during the Depression to erect barriers to the movement and marketing
of milk where the motivations were primarily economic. These took
many forms--refusal to inspect farms or plants beyond the existing
milkshed, outright exclusion by ordinance of milk from plants outside
the city limits or a prescribed radius, excessive fees for inspection
at a distance from the city, and many others. Differences in the
regulations also acted as a barrier in many cases, a circumstance
which the PHS model ordinance and code was designed to correct.
Differences between jurisdictions in the standards of identity of
milk products also operated as barriers to marketing of packaged
milk products--for example, the minimum requirement for milkfat in
whole milk varied substantially, so that whole milk which was quite
legal in the market where it was bottled could not be sold in an
adjoining jurisdiction. (See Taylor et al, 1939.)

Pasteurization was generally required and thus more common in
the larger cities (table 56), although sometimes exceptions were made
for producer-dealers. This had a significant impact on the structure
of fluid milk markets, as many producer-dealers could not afford to
purchase pasteurization equipment when the regulation was changed to
require it. Wide differences in the numbers and importance of
producer-dealers in different markets reflected such differences in
regulations to a considerable degree.

Table 56--Percentage of milk pasteurized in municipalities of 1,000 or more population, by size, 1924 and 1936

Population	1924		1936	
	No. of cities reporting	Percentage of milk pasteurized	No. of cities reporting	Percentage of milk pasteurized
1,000- 10,000	73	33.6	1,766	39.3
10,000- 25,000	105	42.5	281	58.2
25,000-100,000	104	66.8	169	72.6
100,000-500,000	37	81.7	65	85.9
Over 500,000	9	98.1	10	97.5
Total	328	68.6	2,291	74.7

Source: Fuchs and Frank, 1938, p. 29.

The economic impact of the use of sanitary regulation as a barrier to milk movement in the thirties was relatively modest. Since ample milk supplies were generally available in the established milksheds, it resulted in fairly modest price enhancement (see Spencer, 1933, for example). As we will see, such barriers became much more significant during and after World War II, as demand for fluid milk increased much faster than supply.

The imposition of sanitary regulations in a market where it had not previously existed or where enforcement had been lax raised costs sharply for small producers. Many dropped out of milk production or switched to other markets where the requirements were less onerous. The experience in St. Louis provides an excellent example (Harris, 1958, p. 13; Freemyer, 1950).

Before 1935, there had been little enforcement of sanitary regulations in the St. Louis market. Producers could enter the market without farm inspection or registration with the local health department. Public criticism of the quality of the milk supply led to new ordinances in 1934 and 1936, the latter patterned after the model ordinance recommended by the Public Health Service. Perhaps more important, funds were provided to enforce it. Number of producers supplying the market dropped rapidly from 12,000 in 1934 to 10,000 in 1936 after the first efforts and to 5,000 in 1938 (Freemyer, 1950, p. 24). Average receipts per producer nearly doubled.

FEDERAL REGULATION OF FLUID MILK MARKETS

The catastrophe of the Great Depression plunged the entire country into economic chaos. The election of Franklin D. Roosevelt and the coming of the New Deal brought new approaches to the Government role in the economic life of the nation. Roosevelt's first priority was to deal with the bank collapse which had started before he took office. His second priority was agriculture (Schlesinger, 1959, pp. 27-28).

The original intention of Secretary of Agriculture Henry Wallace and his advisers had been to push for enactment of the "agricultural adjustment" or "domestic allotment" plan developed by John D. Black of Harvard University, Beardsley Ruml, and M. L. Wilson just before the crash. It proposed to provide farmers with price supports in exchange for a pledge to reduce production. The "adjustment" emphasis was on persuading farmers to shift to more profitable crops, not just on decreasing overall production.

It was decided to broaden the range of options and to give the Secretary broad discretion to choose among them. The options included marketing agreements, export subsidies, withdrawing land from production by lease, price supports through Government loans and purchases, and licensing of processors to obtain compliance. A bill including these optional programs, packaged with mortgage relief for farmers and a "greenback" program designed to inflate the money supply and raise prices, became law in May 1933 (Schlesinger, 1959, p. 39).

The National Industrial Recovery Act was passed the next month. It placed major emphasis on "agreements" between business and labor, with the government as a party. Many in the administration regarded the marketing agreements provision of the Agricultural Adjustment Act as similar (perhaps identical) to the agreements under the National Industrial Recovery Act. Some felt that all such agreements, regardless of the authorizing legislation, should be administered by General Johnson and the National Recovery Administration. This idea was soon put to rest and authority under the NIRA pertaining to milk and other agricultural products, except for the labor provisions, was transferred to the Agricultural Adjustment Administration. The authority under the NIRA was used but little by AAA (Mehren, 1947, p. 3).

Marketing Agreements

George Peek, the first administrator of the new Agricultural Adjustment Administration, saw marketing agreements backed up by licenses for processors as the first line. They began to be used for milk and specialty crops, but efforts to adopt agreements for sugar and tobacco were turned down within the AAA (Schlesinger, p. 55). 1/ A marketing agreement was an agreement between the Secretary of Agriculture, associations of producers, processors, and handlers setting prices and other terms of trade. Since the agreement was voluntary, not all processors or handlers signed. To ensure compliance throughout the market, all processors or handlers were covered by a blanket license for the entire market.

1/ This and the following section are based largely on Black, 1935, pp. 83-151.

The first marketing agreement was for Chicago. It had been drafted by the Pure Milk Association and the larger dealers and was proposed to Secretary Wallace on the day the Agricultural Adjustment Act was signed. In general, the proposal embodied the provisions of the agreement then in effect in the Chicago market between the cooperative and the major dealers. Both the cooperative and the dealers wanted it because of the proliferation of "chiselers," "price gougers," or "competitors"--the choice of terms depending, of course, on whose ox was being gored at the moment. A major advantage to the dealers was fixing minimum retail prices, as well as minimum producer prices. Retail price floors, of course, had not been found in the agreement between the cooperative and the dealers existing previously in the market.

After considerable revision of the proposal presented by the Chicago group, it was signed and became effective August 1, 1933. In the meantime, requests for marketing agreements were coming in from all over the country. By the end of the year, more than 200 requests had been received from different markets. Only 15 agreements had become effective--

Market	Effective Date
Chicago	August 1
Philadelphia	August 25
Detroit	August 27
Minneapolis-St. Paul	September 2
Baltimore	September 29
Knoxville	October 9
Evansville	October 23
Des Moines	October 25
New Orleans	October 28
Boston	November 3
Alameda County (Oakland)	November 7
Los Angeles	November 17
St. Louis	November 22
San Diego	December 15
Richmond	December 20

The general process was that a proposed agreement was submitted jointly by producer and dealer groups, usually the larger dealers. A conference was then held between the contracting parties and the three offices of the AAA having an interest--the Dairy Section, the Consumers' Counsel, and the General Counsel. Here, attempts were made to adjust differences. Then, a public hearing was held--after the first four months, in the market involved. The AAA then drew up the final agreement which was sent to the Secretary for initialing. That proposed agreement went out to the market for signature of the contracting parties, after which it was signed and became effective.

Violations of the agreements, but particularly of the licenses as they applied to those who had not signed the agreements, were widespread. Most--in some markets, all--of the violations were of the fixed retail prices. The chainstores vigorously opposed retail price fixing, particularly the differentials from home delivery prices, if any. The violations were dealt with inadequately and became a significant problem.

Strong differences of opinion developed both within the AAA and in the industry, principally over these issues...

o Resale prices—whether or not to fix them and at what levels.
o The level of producer prices—the prices in the 1933 marketing agreements had averaged 39 cents per hundredweight above those previously paid, raising prices above a competitive level with prices for manufacturing uses in several markets. Producer groups pressed for even higher prices—closer to parity—while others argued that they should be reduced, since enforcement was made especially difficult by the substantial differential from manufacturing milk values.
o Production control—base-rating plans had been incorporated in most agreements and there was great debate between the losers and the gainers. A major issue was the equalization of sales, a form of pooling of base ratings.
o The checkoff (fee for administration). Opposed by nonmembers of cooperatives and by independent dealers.
o Representation of minority interests on local boards.

In an attempt to deal with these problems, a national conference was held, new policies were developed and announced in January 1934.

Licenses

Under the new policy, all marketing agreements were terminated February 1, 1934, and the licenses continued until replaced by new licenses conforming to the new policy. Licenses would be issued first and then new marketing agreements developed. The main features of the new policy were—

o Producer prices for fluid milk were to be kept in line with prices of competing dairy products. Parity prices would be sought only through a general program of production control for all milk. Two methods were used to determine the Class 1 differential, updating from a 1925-29 base in the local market and the "competitive approach" which embodied the principles of locational economics. In most markets, they gave similar results, a Class 1 differential of about 1 cent per quart at the border of the local milkshed (Black, 1935, p. 118), although in a few markets the historical method gave significantly higher results, likely indicating significant market power on the part of local cooperatives in the 1925-29 base period.
o Base-rating plans would be used only to adjust production seasonally. Restrictions on entry by new producers were not allowed.
o Resale prices would not be regulated, except possibly minimums in some markets to protect distributors against unfair competitive practices.
o Broader representation on the local advisory groups.

Not all of these policies were in fact carried out as 48 new licenses were issued during 1934 and most of the 1933 licenses redrafted. Marketing agreements were seldom concluded in fluid milk markets after 1933. The orders became increasingly a Federal effort, with less decision-making by local groups, although they continued to be active in proposing actions and making their views known.

A production control plan for all milk was developed and offered for consideration, but it was dropped because of considerable opposition to it and doubts about it within the Administration.

The producer prices written into the new licenses did not conform closely to the competitive basis. Strong pressure from producer groups usually led to a price level somewhat above the competitive level. A review of prices in 42 licenses in September 1934 found that, on the average, they were 30-35 cents above the competitive level (Black, 1935, p. 144).

By late 1934, activity in issuing new licenses virtually came to a halt. Only two licenses were issued after December 1, 1934, and one of those replaced an old one. Three existing licenses were substantially overhauled. Nine others were cancelled, mostly because of court decisions holding to a narrow interpretation of the interstate commerce clause.

The Supreme Court decision in the Schechter case involving the National Industrial Recovery Act and other court decisions led to doubt over the constitutionality of the Agricultural Adjustment Act and it was rewritten and passed by Congress in August 1935. It replaced licenses with "orders" to be issued by the Secretary. Recognizing the pronouncement of the Court in the Schechter case and elsewhere that Congress can only delegate its power under specific rules of action, it specifically authorized classified (use) pricing and location price differentials. It also authorized marketwide pooling, requiring that individual handler pooling must be approved by three-quarters of the producers in the market, while two-thirds approval was required for all other matters. This provision was in deference to the opposition of major cooperatives to individual handler pooling. The entire provision for voting by producers, including the authorization for cooperatives to vote for all of their members, was new. It had been proposed by some milk distributors that a vote approving the order be required by both producers and dealers. This "dealer veto" was not included in the bill reported out of committee. It also limited the Act to designated commodities which were "in the current of interstate or foreign commerce or... directly affected such commerce." The latter phrase was the basis for a less-restrictive interpretation of the commerce clause which was adopted by the Supreme Court by the end of the thirties.

The provisions of the 1935 act were reenacted with very little change as the Agricultural Marketing Agreement Act of 1937 and have provided the basis for Federal milk marketing orders since that time. The declared policy of the Act became to establish and maintain such orderly marketing conditions for agricultural commodities in interstate commerce as would provide farmers with parity prices. This Act withstood the constitutional test (see Murphy, 1955, p. 167).

Market Orders

With the passage of the 1935 act, market orders gradually replaced licenses, although licenses and agreements then in effect continued until replaced. The last marketing agreement in the Topeka, Kansas, market was not discontinued until 1947. Most of the others had been replaced by 1941 (table 57).

Table 57--Number of markets regulated by Federal orders, licenses, and marketing agreements, 1933-1941

Year	January 1	During year	December 31		
			Orders	Others	Total
1933	0	15	0	15	15
1934	15	51	0	50	50
1935	50	51	0	30	30
1936	30	31	5	18	23
1937	23	25	7	16	23
1938	23	26	11	14	25
1939	25	28	15	12	27
1940	27	29	20	7	27
1941	27	28	21	5	26

For later years, see table 65.

In 1934, 51 markets were regulated for at least a part of the year and 50 markets in 1935. Thereafter, the numbers declined. The markets regulated in 1934-35 included a substantial number of smaller markets--plus four in California--where the test of interstate commerce at the very least cast considerable doubt on the constitutional basis of regulation (tables 58 and 59). Many of these markets gradually were phased out of Federal regulation. At the same time, a number of States were setting up their own programs, replacing Federal regulations in such markets. Several major markets--New York City and Philadelphia, for example--attempted to make do with State regulation for several years.

By 1940, Federal markets were concentrated in the Northeast and North Central regions, with 26 of 29 markets in those two broad territories--and most of the milk (table 58). The number of markets dropped by nearly one-half, but the amount of milk under regulation increased nearly one-half, primarily because New York and Chicago came under Federal regulation. A substantial share of the smaller markets were no longer regulated.

Table 58--Federal orders, agreements and licenses, by region, 1933-1940

Region	:	Number of markets during year			:	Producer deliveries	
	:	1933 :	1935 :	1940	:	1935 :	1940 1/
	:				:	Bil.	lbs.
New England	:	1	5	4		1.1	1.2
Middle Atlantic	:	2	2	2		1.4	5.1
South Atlantic	:	1	3	0		1.4	0.0
East North Central:	:						
Eastern group	:	1	11	4		.7	.4
Western group	:	2	7	4		.4	1.8
West North Central:	:						
Northern group	:	2	7	6		.7	.8
Southern group	:	0	5	6		.5	.6
East South Central	:	1	0	0		0.0	0.0
West South Central:	:						
Northern group	:	0	2	0		*	0.0
Southern group	:	2	2	1		*	.1
Mountain	:	0	3	1		.1	.1
Pacific	:	3	3	1		.4	.1
Total	:	15	50	29		6.8	10.3

*Less than .05 billion pounds.
For later years, see tables 66 and 67. Regions as in figure 4.
1/ Year ending June 1940.

Table 59--Federal orders, licenses and agreements, by size of market, 1935 and 1940

Annual producer deliveries (million pounds)	:	Number of markets		
	:	1935	:	1940
Less than 25	:	19		6
25-49	:	10		6
50-99	:	7		4
100-399	:	6		9
400-599	:	3		1
600-999	:	2		0
1,000 or more	:	3		3
Total	:	50		29

For later years, see table 71.

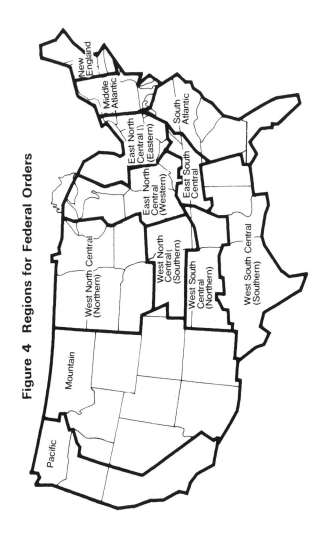

Figure 4 Regions for Federal Orders

Changes in legislation and administrative policy between 1933 and 1935 meant that the Federal milk order program developed along considerably different lines than those of the early years. Gone were the efforts to fix resale prices and margins which had provided the incentives for processor acceptance of the marketing agreements. The role of cooperatives changed significantly, as their participation was a key element in the Federal presence in fluid milk markets. The orginal emphasis on rapidly raising prices to parity levels was replaced with a competitive model of what prices would be like in a freely operating market, but altered by substantial modifications necessary to obtain acceptance by producers.

In practice, this led to compromises between the competitive model of what relationships should be and historical relationships. In particular, the issues were most often between groups of producers within a market who had received higher prices historically and those who had not. These advantages arose out of individual handler or less-than-marketwide pooling in the past. Nearby differentials were incorporated in some orders to approximately equal the price advantages that producers close to the market had previously enjoyed.

STATE MILK CONTROL

The economic distress of dairy farmers caused by the Depression led to efforts by States as well as the Federal government to take actions which would assist farmers. The Constitutional authority of the Federal government to deal with economic questions was much more narrowly construed at the beginning of the New Deal than it was by the end of the thirties. Since nearly all fluid milk markets were local in character and many were entirely within one State—including both market and milkshed—, there was a question as to whether Federal marketing agreements and licenses could be used. Early court decisions found that Federal regulation of purely intrastate markets was not legal. All of this provided incentive for milk control by the States.

The police powers of the State gave clearcut authority in most States to regulate milk marketing directly. Thus, the Federal devices of marketing agreements and licenses and later market orders were not required. The State could and did issue regulations specifying minimum or maximum prices and other terms of trade.

The first State milk control law was passed in Wisconsin in 1932 and regulations issued under it for the Milwaukee market in December, but resale price regulations were not effective until February 1933. The first act was not very effective and it was revised during 1933 and licenses issued for 18 of 27 markets which were subject to it. Another 50 small cities had come under voluntary agreements which did not raise prices but were supposed to prohibit price-cutting (Black, 1935, p. 337).

New York enacted a State milk control law in April 1933, growing out of an intensive study of the milk problem by a joint legislative committee which recommended legislation (Joint Legislative Comm., 1933). Northeastern States as far south as Virginia adopted milk control laws in 1933-35, except for Delaware where a bill was introduced but did not pass the legislature. The Maryland law enacted in

1935 never became effective and was declared unconstitutional in 1936 but a new law was passed the following year. While many of the original acts were effective for one to four years, being considered emergency acts, nearly all were reenacted in modified form and lasted for twenty years or more (Spencer and Christensen, 1954, pp. 8-9).

In the midwest, Ohio, Michigan, Wisconsin, and Indiana had State milk control for some time during the thirties, but all had phased out by 1941.

In the South, Georgia, Florida, and Alabama enacted State milk control laws during the thirties which continued to be used for at least thirty years. Alabama's law is still on the books but is unused, a Federal order having been introduced in 1982. Louisiana and Texas had brief spells during the thirties but they did not persist in the forties.

Five western States passed milk control acts during the thirties. Those in Montana, Oregon and California were long-lived, but Utah and Washington (where the legislation was part of a general agricultural adjustment act) were gone by 1941. California still sets producer prices.

State milk control took a wide variety of shapes and forms. Many but not all included resale price regulation, which helped to make them popular with milk dealers. A number relied rather heavily on restrictive licensing of milk dealers to control entry of new firms into the market. Classified pricing was used by most but not all States, but marketwide pooling was uncommon.

EFFECTS OF REGULATION

Both Federal and State programs were designed to raise producer prices and did so, at least modestly. From December 1932 to December 1934, the increase in the quoted Class 1 price to producers was—

    25 percent in 17 markets under Federal control
    37 percent in 13 markets under State control
    20 percent in 10 markets not under control

Black (1935, p. 344) emphasizes that these are nominal prices only and no one knew how much undercutting of prices occurred in either period. Another tabulation, starting from May 1933 when prices were even lower and including more markets, showed somewhat larger rises (table 60).

The net effect of Federal or State control on returns to producers depends upon at least three factors:

o The increase in the announced Class 1 price.
o Requiring dealers formerly buying on a flat price basis to pay the full Class 1 price.
o The decrease in producer returns caused by increased supplies and larger surpluses, attracted by higher Class 1 prices.

Spencer and Christensen estimated the effects of these three factors on New York producers from 1933 to 1953 when they were under State

Table 60--Estimated price increases for Class 1 milk,
May 1933-September 1934

Markets and type of control	Price May 1933 1/	Increase by--			
		May 1934		September 1934	
	Dollars	Cents	Percent	Cents	Percent
23 markets under Federal control	1.45	42	28	71	49
11 markets under State control	1.73	54	31	77	44
16 markets not under control	1.66	26	16	35	21

1/ Price per hundredweight for 3.5 percent milk, f.o.b. city.

Source: Black, 1935, p. 344, from W. A. Wentworth, Dairy Industry Committee.

control only for four years and then under joint Federal-State control (table 61). The increase in producer returns was about 15 percent in some periods during the thirties and 4-6 percent in the late forties and early fifties.

PROGRAMS FOR MANUFACTURING MILK AND BUTTERFAT

Upon recommendation of a national Dairy Advisory Committee, the Federal Farm Board, on January 9, 1930, made a loan to Land O'Lakes Creameries, Inc., to enable the cooperative to withhold temporarily some of its own butter and, if necessary, to purchase additional butter on the open market in order to stabilize prices. The cooperative offered to buy at market quotations whenever prices of butter were 35 cents per pound or lower, but no butter was offered to it. 1/
By March 15, it had accumulated about 5 million pounds of butter from its own production which was sold back to the trade by May, the start of the heavy production season, at a profit. Thus ended the first price stabilization experiment in the marketing of manufactured dairy products, conducted under authority of the 1929 Act.
The Agricultural Adjustment Act of 1933 provided three ways for improving prices and income to dairy farmers.

1/ This section is based on Rojko, 1957, pp. 149-155. See also Black, 1935; Lininger, 1935; and Nicholls, 1939, p. 342.

Table 61--Estimated effects of State and Federal milk control programs on the net farm price of milk in New York State, 1933-1953

Period	Number of months	Type of control in effect	Increase in producer price due to:			Decrease in producer price due to lower utilization	Net gain in price paid producers	
			Higher Class 1 price	Adjustment of flat price	Total		Amount	Percent
			------Cents------					
July-December 1933	6	State	19	5	24	--	24	15
January-December 1934	12	State	18	7	25	--	25	15
January 1935-March 1937	27	State	13	6	19	--	19	10
September 1938-January 1939	5	State and Federal	20	8	28	2	26	15
July 1939-May 1942	35	State and Federal	8	7	15	4	11	5
July 1946-June 1948	24	State and Federal	--	3	3	1	2	--
July 1948-December 1950	30	State and Federal	27	9	36	9	27	6
January 1951-December 1953	36	State and Federal	20	7	27	11	16	4
Average for all periods of control	175		14.8	6.6	21.4	4.8	16.6	5.2

Source: Spencer and Christensen, 1954, p. 20.

o Milk and its products were designated as basic commodities, authorizing the same price-support and production-adjustment operations as for storable commodities. "A dairy-adjustment program was presented to producers, but the support it received from the dairy industry was not deemed sufficient to warrant its adoption." (U.S. Agric. Adj. Admin., 1935).

o Marketing agreements, licenses, and Secretary's orders.

o Authority to use funds for the expansion of markets and disposal of surplus agricultural products.

Under this and similar later legislation, some support for prices for milk and butterfat were carried out from mid-1933 until early in World War II. The Department did not announce specific price support levels or specific purchase prices. Purchases generally were made on the basis of competitive bids and the quantities purchased usually did not exceed those that could be used for school lunch, institutional, and welfare purposes. Determinants of the quantity and kind of product purchased under these early programs were:

(1) Effectiveness of producer groups in pressing for assistance.

(2) Suitability of the product in meeting the food requirements of those on the relief rolls.

(3) Ability to make an observable improvement in the market price with the funds available. (U.S. Senate, 1961, p. 5).

With few exceptions, the early purchase programs were carried out with wide discretionary powers from the Administrators of the Agricultural Adjustment and Federal Emergency Relief Administrations. The quantities purchased mainly for price-support operations are shown in table 62.

In the summer of 1933, the Secretary of Agriculture authorized Land O'Lakes Creameries, Inc., to purchase surplus butter for resale to the Administration. Between August 17 and October 25, the cooperative bought 11 million pounds which was donated to the Federal Surplus Relief Corporation for relief purposes with the understanding that the Relief Administration would also expend some of its own funds for the purchase of butter and cheese.

The Federal Surplus Relief Corporation (FSRC) was chartered under the laws of the State of Delaware in October 1933 for the purpose of purchasing and processing commodities for relief distribution. In November 1935, the charter was amended to call it the Federal Surplus Commodities Corporation and to change the membership so that the direction of the corporation was transferred from the Federal Emergency Relief Administration to the Department of Agriculture. This brought a shift in emphasis from relief to removing agricultural surpluses and encouraging consumption. Dairy products distributed were obtained by (1) direct purchases with the corporation's own funds, (2) donations from the Agricultural Adjustment Administration, and (3) donations from State Emergency Relief Administrations.

Table 62--Dairy products purchased by the U.S. Department of Agriculture, mainly for price support, 1933-41

Year	Butter 1/	Cheese 2/	Evaporated milk	Whole milk equivalent, all purchases		Nonfat dry milk		
				Quantity purchased	Purchases as a percentage of production of milk on farms	Quantity purchased	Purchases as a percentage of-- Production of nonfat dry milk for human use	Purchases as a percentage of-- Total solids-not-fat produced on farms
	1,000 pounds	1,000 pounds	1,000 pounds	Million pounds	Percent	1,000 pounds	Percent	Percent
1933	3/43,234	--	--	869	0.8	--	--	--
1934	4/24,624	5/17,936	6/400	675	.7	--	--	--
1935	7,055	192	47,027	244	.2	15,840	8.4	0.2
1936	2,951	932	6,160	82	.1	3,594	1.6	7/
1937	8/3,049	9/138	19,636	104	.1	23,188	9.5	.3
1938	141,979	3,463	10/19,470	2,916	2.8	11/31,260	10.8	.3
1939	25,398	--	3,209	515	.5	5,035	1.9	.1
1940	10,604	4,354	65,903	397	.4	12/7,317	2.3	.1
1941	11,454	--	4,350	238	.2	12/2,742	.7	7/

1/ Includes 132,006,000 pounds purchased by Dairy Products Marketing Association during 1938-41. 2/ American cheese unless otherwise specified. 3/ Includes 11,051,046 pounds purchased by Land O'Lakes prior to mid-October 1933. 4/ Includes 5,908,020 pounds purchased with Federal Surplus Commodities Corporation funds in 1934. 5/ Includes Swiss cheese purchased in August. 6/ Purchased by F.S.C.C. during 1934. 7/ Less than 0.05 percent. 8/ Purchased by F.S.C.C. with State funds. 9/ Purchased by F.S.C.C. under State programs for flood relief. 9/ Purchased by F.S.C.C. with State funds. 10/ Includes 435,000 lbs. bought with State funds by F.S.C.C. in September and October and 19,035,000 pounds acquired by F.S.C.C. in November and December in exchange for fluid milk under the New York milk diversion program. 11/ Includes 1,001,000 pounds acquired by F.S.C.C. in November in exchange for fluid milk under the New York milk diversion program. 12/ Includes 2,336,000 pounds in 1940 and 2,742,000 pounds in 1941 acquired for relief distribution by the Surplus Marketing Administration from D.P.M.A.

Source: Rojko, 1957, pp. 152-153. Compiled from records of operating agencies.

In October 1933, the Dairy Marketing Corporation was formed to handle purchases of surplus dairy products. The stockholders of the Corporation were the National Cooperative Milk Producers' Federation, the American Association of Creamery Butter Manufacturers, the International Milk Dealers' Association, and the National Cheese Institute. Purchases were to be made only upon instructions of the Secretary of Agriculture. The Department agreed to take over all products acquired. From October until December 16, when its agreement with the Department of Agriculture terminated, the Corporation purchased 32 million pounds of butter, which was turned over to the FSRC.

In December 1933, the FSRC began to purchase butter and cheese through bids. Direct market purchases in December 1933 and early 1934 included 46 million pounds of butter and 6 million pounds of cheese, financed from Treasury advances of $11 million in anticipation of processing taxes on dairy products.

In 1934, funds for the purchase of surplus dairy products were made available under the Jones-Connolly Act of April 7, 1934. Section 37 of the Agricultural Act of August 24, 1935, provided additional funds that could be used to purchase surplus dairy products. From 1934 to 1938, the Government spent $22 million from funds provided under these Acts and from Federal and State Emergency Relief funds in direct market purchases and for relief distribution of the following dairy products:

	Million pounds
Butter	35
Cheese	13
Evaporated milk	88
Condensed milk	1
Nonfat dry milk	50

Section 32 of the Agricultural Act of 1935 authorized the Secretary of Agriculture to use to 30 percent of the annual customs receipts to encourage exports and domestic consumption of commodities by diverting them from normal channels of trade or by increasing their use among low-income groups. Beginning in 1937 for butter and evaporated milk and in 1938 for cheese and nonfat dry milk, direct market purchases of dairy products to provide price assistance to dairy farmers by removing surplus commodities from normal trade channels were financed chiefly from these funds. These products were disposed of through relief distribution. In 1937-41, $57 million of section 32 funds was spent in purchases for surplus removal:

	Million pounds
Butter	176
Cheese	8
Evaporated milk	78
Nonfat dry milk	28

All the quantities purchased by the FSCC were obtained directly
in the marketplace, except for some nonfat dry milk and 118 million
pounds of butter which were obtained from the Dairy Products Marketing Association.

The Dairy Products Marketing Association (DPMA), a non-profit
organization with a membership of eight regional butter marketing
cooperatives, was set up in 1938 to help operate the Government stabilization program for butter. Loans were made to the Association
by the Commodity Credit Corporation (CCC) to buy butter at specified
prices. Butter, cheese, and nonfat dry milk bought by the Association was held in storage for possible resale through commercial channels at prices which covered the purchase price plus handling and
carrying charges. Dairy products not resold to the trade could be
sold to the Surplus Marketing Administration for relief distribution.
DPMA, acting as an agent for USDA, also purchased dairy products during World II for Lend-Lease and Government use. The last purchases
of DPMA occurred in the spring and summer of 1947. Under the price
stabilization program, DPMA bought 132 million pounds of butter in
1938-41, 114 million pounds in 1938 alone.

Government purchases of dairy products for price support during
1933-41 had fairly minor effects on the overall price structure for
dairy products in most years. But since the products were bought
during periods of abnormally low prices, even relatively small purchases could substantially affect the market price of a dairy product
in the short run. For example, when Land O'Lakes bought 11 million
pounds of butter on the Chicago and New York markets between August
and October 1933, the price of butter, which had dropped to 18 cents
a pound before purchases began, increased to 23 cents in Chicago and
approximately 24 cents in New York within a few days. Total pre-
World War II purchases were less than 1 percent of total production
of milk in any year except 1938, when they were close to 3 percent
(table 62). The average price received by farmers for milk might
have been about 6 percent lower in 1938 if no purchases had been
made, assuming that quantities distributed for relief replaced little
in the way of market purchases.

Even though there may have been some substitution for regular
market purchases, Government purchases for price support, distributed
later for relief, undoubtedly tended to increase total consumption of
dairy products, and Government expenditures probably were chiefly a
net addition to the income of dairy farmers. The remaining smaller
supplies in commercial channels normally would sell for more total
dollars than the larger supply because of the relatively inelastic
demand for milk and most manufactured dairy products.

FOOD DISTRIBUTION PROGRAMS [1]/

Some of the food distribution programs discussed in this section
had their origin in attempts to improve prices and income received by

---

[1]/ See also U.S. Fed. Surplus Commodities Corp.; U.S. Surplus
Mktg. Admin., 1941; Foote, 1941 and 1947; Gold et al, 1940; U.S.
Agric. Mktg. Admin., 1942; U.S. War Food Admin., 1943 and 1944b.

farmers during periods when agricultural commodities were in surplus supply. The immediate objective of these programs frequently was to supplement price support operations by providing outlets for surpluses acquired under the support program and by increasing consumption through their own purchases. Some programs, such as special school milk, were designed to supplement price supports for manufacturing milk and butterfat by increasing consumption in fluid outlets and thereby reducing the quantity of milk available for the production of surplus manufactured dairy products bought by CCC. Although the immediate objectives of most of these programs were to give price assistance to dairy farmers and to provide outlets for dairy products acquired under price support programs, their long-run objectives were to bring about consumption levels that would improve the health and well-being of the nation and to expand consumption of agricultural commodities and other food. The price effect of these programs depends on the scale of operation which, in any given year, may be affected by the size of surplus supply.

In direct distribution programs, the Federal Government purchased food for donation to schools, charitable institutions, and welfare groups. These programs have been discussed in conjunction with the Federal Surplus Commodities Corporation (FSCC) purchases prior to World War II. Quantity distributed varied, depending on the price and surplus supply position (table 63).

In October 1937, an experimental milk distribution program for needy people was begun in Boston, continuing until August 1939. FSCC purchased raw milk, paid processing costs of about 2 cents per quart, and distributed the milk free to people on direct relief and other eligibles for 2 cents per quart. Between 1939 and 1942, a relief milk program was in operation in six cities--Boston, Chicago, New Orleans, New York City, St. Louis, and Washington, D.C. The elements of the program included: (1) farmers received a price for milk above the surplus milk price but below the Class 1 price; (2) the recipients paid about 5 cents per quart (4 to 6 cents); and (3) the Surplus Marketing Administration paid for the remaining cost of the milk, amounting to about 2 cents. Significant increases in the consumption of fluid milk occurred among participants in the programs, although the increase included some replacement of evaporated milk (Stiebeling et al, 1942, and Sullivan, 1942).

The school milk program was started on an experimental basis in May 1940 and was merged with the school lunch program in 1943. In 1941-42, about 52 million pounds of milk were distributed at a cost of $1.5 million from Section 32 funds. Expenditures in 1943 were $4.5 million. In the operation of this program, the school made agreements with the local dairy and the Agricultural Marketing Administration. AMA reimbursed the local agency in an amount equal to cost of Class 1 unprocessed milk. The local agency assumed handling costs and distributed the milk at a cost to school children of not more than a penny a half pint.

The food stamp plan, which was started in May 1939 and discontinued in February 1943, was a Federal subsidy that provided additional food to low-income families. These families were given free

Table 63--Dairy products: Volume and expenditures on products handled under sec. 32, fiscal years 1937-44 1/

Fiscal year	Butter								Cheese		Evaporated milk	
	Food stamp plan		Direct distribution		Exportation		Total		Direct distribution 2/		Direct distribution 2/	
	Quantity	Value	Quantity	Value	Quantity	Value	Quantity	Value	Quantity	Value	Quantity	Value
	Thous. lbs.	Thous. dols.	Thous. lbs.	Thous. dols.	Thous. lbs.	Thous. dols.	Thous. lbs.	Thous. dols.	Thous. lbs.	Thous. dols.	Thous. lbs.	Thous. dols.
1937	--	--	357	122	--	--	357	122	--	--	4,350	246
1938	--	--	15,030	4,129	--	--	15,030	4,129	3,463	499	--	--
1939	143	39	122,287	34,704	802	24	123,232	34,768	--	--	3,202	172
1940	9,588	3,010	31,687	9,408	239	9	41,514	12,427	4,284	642	2/67,814	2/3,628
1941	29,154	10,272	6,619	2,075	--	--	35,773	12,347	--	--	4,350	375
1942	25,105	10,323	9,041	3,221	--	--	34,146	13,544	5,000	1,250	119,494	9,622
1943	5,872	2,544	1,366	520	--	--	7,238	3,064	1,999	611	5/	5/
1944	--	--	--	--	--	--	--	--	--	--	32,340	3,611

Fiscal year	Nonfat dry milk		Fluid milk						Total	
	Direct distribution 2/		Direct distribution		Diversion		Total		Total	
	Quantity	Value	Quantity	Value	Quantity	Value	Quantity	Value	Quantity (milk equivalent) 3/	Value 4/
	Thous. lbs.	Thous. dols.	Thous. lbs.	Thous. dols.	Thous. lbs.	Thous. dols.	Thous. lbs.	Thous. dols.	Thous. lbs.	Thous. dols.
1937	--	--	--	--	--	--	--	--	16	368
1938	6,471	367	26,870	881	--	--	26,870	881	362	5,876
1939	13,997	707	131,722	3,349	--	--	131,722	3,349	2,603	38,996
1940	2/5,006	2/347	4,680	148	80,247	693	84,927	841	1,104	17,885
1941	2,120	23	--	--	174,935	2,111	174,935	2,111	900	14,856
1942	7,300	1,095	--	--	192,085	3,869	192,085	3,869	1,182	29,380
1943	985	134	--	--	224,586	6,573	224,586	6,573	389	10,383
1944	--	--	--	--	--	--	--	--	70	3,611

1/ Sec. 32 of the Agricultural Act of Aug. 24, 1935, as later amended, authorized the Secretary of Agriculture to use an amount equal to 30 percent of the annual customs receipts to encourage the exportation of agricultural commodities and the products thereof and to encourage domestic consumption of such commodities by diverting them from normal channels of trade or increasing their use among persons in low-income groups. 2/ Includes the following exportations in 1940: 1,911,215 pounds of evaporated milk valued at $110,851 and 25,000 pounds of nonfat milk valued at $1,575. 3/ Fat solids basis. 4/ Computed from unrounded figures. 5/ Values less than $500.

Source: Letter from Secretary of Agriculture..Methods of Production Control and..Price Support. 89th Cong., 1st Sess., House Doc. No. 57, 1955.

blue stamps to be used for the purchase of specified surplus foods on condition that they buy a certain quantity of orange-colored stamps that could be used for the purchase of any food. Butter was the only dairy product listed as a surplus food eligible for the program and was so designated three-fourths of the time.

MARKETING AGREEMENTS FOR MANUFACTURED PRODUCTS

The Agricultural Adjustment Act of 1933 authorized marketing agreements for essentially all agricultural products. An attempt was made in 1933 to develop a marketing agreement for butter, but disagreements between various factions in the industry were too great. Interestingly, the first draft of the proposed agreement provided for fixing a weekly price for butter by a butter board of five members, representing butter manufacturers and cooperatives. This would have replaced the system of establishing the base butter price in exchange trading which has been the subject of much study and condemnation since that time, although the system still operates.

A marketing agreement for evaporated milk became effective in September 1933. It established producer prices in six regions at a differential from butter prices in the nearest market—both butter and cheese prices in the Midwest and New York. Wholesale prices were uniform throughout the country except for a few differentials, with a 15-cent-per-case spread between maximum and minimum, primarily to provide for a differential between advertised and unadvertised brands.

Manufacturers producing about 95 percent of the total volume signed the agreement. The original marketing agreement was replaced June 1, 1935, by a marketing agreement and license which discontinued fixing wholesale prices by formula and substituted price filing by manufacturers. Minimum prices to producers were based on butter and cheese prices (U.S. Agric. Adj. Admin., 1937a, p. 73).

Part of the original motivation for the marketing agreement had come from changing competitive conditions and market shares in the early thirties. The six "standard brand" companies—the larger companies which sold nationally advertised brands—suffered a loss from 58.5 percent of the total market in 1931 to 52.2 percent in 1932. They sharply increased their sales of "off-brand" evaporated milk at lower prices from 8 percent of their sales in 1931 to 14 percent in 1932 and 16 percent in 1933. The marketing agreement and, especially, the license stabilized competitive conditions in the industry and the standard brand companies regained their market share by the last half of the thirties (Baker and Froker, 1945, pp. 63-65).

A dry skim milk agreement also went into effect in September 1933. Nearly all manufacturers signed the agreement. It provided for wholesale price filing and adherence to announced prices. Producer prices were not set, since skim milk was largely a by-product of butter production.

The dry skim milk agreement continued until 1941 and the evaporated milk agreement until 1947. The principal effect of both was to somewhat restrain price cutting (known to its opponents as "chiseling" or "gouging") in the processors' market. The evaporated milk agreement also had a similar effect in the producer market, although it did not attempt to raise the general level of prices.

IN SUMMARY

The massive collapse of the nation's economy during the Depression brought Government intervention in markets for many products at Federal, State, and local levels in an attempt to deal with the disastrous economic conditions. Local sanitary regulations were widely used to erect barriers to the movement of milk in an effort to help local producers. Both State and Federal governments intervened to set fluid milk prices, applying the principles of classified pricing which had been developed by the cooperatives and making them effective for all milk. The Federal government also gave some assistance to producers of milk for manufactured products by puchasing those products for use in relief programs.

10
Changes in the Public Role—
World War II and Beyond

Public regulatory and assistance programs for the dairy industry changed significantly in the four decades since the beginning of World War II. Many of these modifications were in response to the changes in the dairy industry and the general economy. There was evolution both in the dairy industry and in the public programs and they are interrelated. Many times, changes in the industry preceded changes in public programs, but at other times the administrators of Federal programs were leading the industry.

SANITARY REGULATION

Substantial barriers to intermarket movement of milk erected in the thirties by sanitary regulation and its administration continued into the forties. In the mid-fifties, approximately a quarter of 102 markets in the Midwest required special provisions to bring in outside milk (Williams et al, 1970, p. 13). The extent of these restrictions throughout the country was documented in 1955 (U.S. Agric. Mktg. Serv., 1955, pp. 20-31).

Attacks were mounted in the courts on barriers to milk movement erected by sanitary regulations. Health regulations imposing restrictions on inspection or licensing that were not reasonably related to sanitary or quality considerations generally were held to be illegal. Regulations excluding milk produced or pasteurized beyond specified distances from the market were often found to be illegal, as were refusals by health authorities to license otherwise pure and wholesome milk merely because the community was already adequately supplied, or because inspection would be too expensive, or for other economic reasons not related to health and sanitation. Similarly, courts held that fees must be reasonable. Lawsuits of this type were brought both in State courts and in Federal courts where interstate commerce was affected.

During the fifties and sixties, court and legislative action and changing attitudes on the part of administrators gradually dismantled trade barriers erected by sanitary regulation. By the mid-sixties, many if not most such barriers had been removed. Increasingly, regulation and inspection of farm suppliers was carried out by State rather than municipal agencies. Reciprocity between States is now much more general, so the cost of duplicate inspection is less (see Jones, 1970). An occasional municipal ordinance still requires local pasteurization of milk, but these are not enforced, recognizing that a court test would invalidate them.

PUBLIC REGULATION OF MILK PRICES

Producer prices of nearly all fluid grade milk are currently regulated by either Federal or State agencies (table 64). Federal orders covered 81 percent of fluid grade milk in 1982 and State regulations 17 percent. More than 90 percent of all milk has been regulated since the mid-sixties, following a fairly steady increase since 1940. The increase has been in Federal order regulation, as State regulation has declined from a high of 45 percent in 1935 to 17 percent in 1982.

Table 64--Federal and State regulation of producer prices for fluid grade milk

Year	:	Percent of fluid grade milk regulated by--		
	:	Federal orders 2/	: States 3/	: Unregulated
1930	:	0	0	100
1935	:	22	45	33
1940	:	35	43	22
1945	:	35	23	42
1950	:	41	24	34
1955	:	51	24	25
1960	:	64	21	15
1965	:	70	22	8
1970	:	80	18	2
1975	:	78	18	4
1982	:	81	17	2

1/ Estimates of the total amount of fluid grade milk before 1950 are rough approximations by the author. All 1935 figures are indicative, not precise.
2/ Includes licenses and marketing agreements.
3/ Excludes concurrent regulation by Federal and State orders.

Federal Orders

In 1942, there were 23 Federal orders and 5 marketing agreements and licenses (table 65). The number of marketing agreements and licenses dropped sharply during World War II, as most of them were replaced by Federal orders. The final marketing agreement in Topeka, Kansas, was replaced by an order January 1, 1948.

Table 65--Number of markets regulated by Federal orders, licenses,
and marketing agreements, 1942-82

Year	:	January	:	During year	:	December
	:		:		:	
1942	:	26		28		28
1945	:	25		28		28
1950	:	35		39		39
1955	:	53		63		63
1960	:	77		80		80
1965	:	76		77		73
1970	:	68		69		62
1975	:	61		62		56
1980	:	47		47		47
1982	:	48		49		49
	:					

For earlier data, see table 57.

The number of markets regulated declined slightly during World
War II, held fairly steady until 1949, and then increased rapidly
during the fifties and early sixties. Numbers decreased thereafter,
as the number of new markets becoming regulated was more than offset
by mergers of existing orders. Mergers picked up in the late sixties
and again since 1975.

In 1945, Federal orders were found mostly in the Northeast and
North Central regions, the principal dairy areas of the country
(tables 66 and 67). Only New Orleans was outside this area. Six of
the 28 markets with 64 percent of the milk were in the Northeast.
Markets in the South Central area came under Federal orders by 1950
and in the Mountain and Pacific regions by 1955. Now, 25 percent of
the Federal order milk is in the three Northeastern markets, 31 percent in the 9 East North Central markets, and 9 percent in the 9
West North Central markets. The 17 Southern markets have 15 percent
of the milk and the 10 Western markets 9 percent.

Federal order markets have grown larger over the years, partly
because larger markets became regulated and more milk became associated with existing markets, but mostly because of mergers among markets (table 68). In 1950, only 10 percent of Federal order markets
had annual producer deliveries of 1 billion pounds or more. In 1980,
40 percent of all markets were in this size class.

In the thirties and early forties, fluid milk markets were
mostly self-contained and supplies of milk were ample. Class 1
utilization was 52 percent of producer deliveries in 1935 (in 15 of
50 Federal license markets for which we have data) and, in 1940, 47
percent. Supplies tightened up after the war and 65 percent of all
producer milk was used in Class 1 in 1947-48. Between 1953 and 1972,
the percentage varied from 59 to 65 percent, depending upon the stage
of the milk production cycle. Since 1974, it has been dropping and

Table 66--Number of Federal milk orders and agreements, by region, 1945-1982

Region	:	Number of markets during the year						
	:	1945 :	1950 :	1955 :	1960 :	1965 :	1970 :	1982
New England	:	3	5	5	5	2	2	1
Middle Atlantic	:	3	2	2	5	4	4	2
South Atlantic	:	0	0	0	1	1	4	4
East North Central	:	11	14	20	24	24	17	9
West North Central	:	10	10	17	19	18	15	9
East South Central	:	0	4	5	5	6	5	5
West South Central	:	1	4	12	15	16	14	9
Mountain	:	0	0	1	4	6	5	6
Pacific	:	0	0	1	2	2	3	4
Total	:	28	39	63	80	77	69	49

For earlier years, see table 58. Regions as in figure 4.

Table 67--Producer deliveries under Federal milk orders and agreements, 1945-1981

Region	:	Producer deliveries (bil. lbs.) in --						
	:	1945 :	1950 :	1955 :	1960 :	1965 :	1970 :	1981
New England	:	1.4	1.9	2.2	4.0	4.4	4.8	5.1
Middle Atlantic	:	7.3	8.0	9.4	13.9	15.5	14.9	16.9
South Atlantic	:	0	0	0	.5	.5	2.7	3.9
East North Central	:	3.7	6.1	10.4	14.9	19.4	22.1	27.7
West North Central	:	1.0	2.0	3.3	4.8	5.7	8.6	16.9
East South Central	:	0	.3	.7	.9	1.3	1.5	2.4
West South Central	:	.2	.3	2.2	3.6	4.3	5.6	7.3
Mountain	:	0	0	*	1.0	2.0	2.2	3.7
Pacific	:	0	0	.8	1.2	1.5	2.7	4.0
Total	:	13.7	18.7	28.9	44.8	54.4	65.1	88.0

* Less than 0.5 billion pounds.
See table 58 for earlier years. Regions as in figure 4.

was 46 percent in 1981.

A number of factors contributed to the decline in the Class 1 utilization percentage, in addition to the stage of the milk production cycle. Nearly all fluid grade milk outside of California is

Table 68--Federal order markets, by size of market

Annual producer deliveries	Number of markets			
	1950	1960	1970	1980
(Million pounds)				
Less than 25	2	0	1	1
25-49	7	6	2	2
50-99	5	11	3	5
100-299	18	41	22	7
400-599	1	8	10	4
600-999	2	6	7	9
1,000 or more	4	8	17	19
Total	39	80	62	47

See table 59 for earlier years.

now under Federal orders. Producers of manufacturing grade milk have been converting to fluid grade at a rapid rate since World War II and much of the increase has become associated with Federal order pools. This is due in part to the adoption of the view that conversion to a single grade of milk is well-nigh inevitable and is also desirable and Federal orders should facilitate it.

Pooling Producer Returns. Pooling of receipts from all handlers and their distribution to individual producers under uniform rules were designed to provide equality of treatment to producers. The legislation also explicitly permits cooperatives to blend the net proceeds of all operations over all members...the "reblending" privilege. Thus, a cooperative with members in more than one Federal order market or with some members not delivering to a Federal order market can provide for uniform methods of distribution among its members. 1/

Two types of pools have been widely used in Federal order markets...marketwide pools and individual handler pools. Under individual handler pools, the returns for milk in each class from the handler are "blended" and distributed to those producers who deliver to that particular handler. Under such circumstances, producers delivering milk to different handlers in a given market will receive different prices for their milk because of differences in utilization between handlers.

Such differences can give some handlers a strong competitive edge in the form of higher blend prices. These handlers can attract preferred larger-volume producers or those whose seasonal production

---

1/ This section draws heavily on Nourse et al, 1962, pp. 44-47.

patterns are closer to their own needs. The individual handler pool creates strong pressure to keep Class 1 utilization high. The other side of this feature is that there is likely to be "homeless" milk.

Thus, while individual handler pooling encouraged handlers to exercise a strong degree of supply control within the local fluid milk market, it also created situations of excess supplies looking for a home. A cooperative often took on the function of finding a home for its members' milk that was not wanted by handlers. Since the outlet for such milk was inevitably in lower-valued uses (usually butter), the cooperative was at a severe disadvantage.

Since, even in the best-managed markets, such conditions were almost certain to arise from time to time, cooperatives handling a major part of the surplus for the market were strongly compelled to seek marketwide pooling. The advent of Federal regulation made it possible to have truly marketwide pooling, in contrast to the cooperative pools of earlier years. All producers were paid under uniform rules with the principal differences being due to butterfat content and location...those farther from the city received less than those close to the city because it cost more to haul their milk to market. (There were also nearby differentials in some markets, which we will come to later.)

The introduction of equalization of producer returns through marketwide pooling altered competitive relations among handlers significantly. Handlers with high Class 1 utilization were no longer able to exercise influence over producers through the high blend prices they were able to pay. Those handlers with lower Class 1 utilization were no longer at a disadvantage. Marketwide pooling reduced producer identification with individual handlers and minimized incentives for making "deals" between producer and handler at less-than-order prices.

Marketwide pooling also provided distinct advantages to cooperatives who were handling the majority of milk in the market. They were no longer disadvantaged as compared with other cooperatives or with nonmember producers. Thus, marketwide pooling was extremely important to the growth of cooperatives in individual markets. It was a necessary condition to the achievement of the kind of economies of large-scale management of milk supplies within a market described in Chapter 5.

Until the mid-fifties, 26-27 percent of all Federal order markets had individual handler pools. These tended to be smaller markets (table 69). In the sixties, the importance of individual handler pools dropped sharply and, in 1970, the last large market with an individual handler pool--Philadelphia--switched to a marketwide pool when it was merged with Baltimore and Washington into the Middle Atlantic order.

Price Formulas. The original marketing agreements, licenses, and orders had specified Class 1 prices in dollars-and-cents terms. Adjustments were made, when needed because of changed economic conditions, by the hearing process. Changes were slow enough during the Depression so that this system was adequate. But, with the advent of World War II in Europe and especially after U.S. entry into the war, conditions changed much more rapidly and requests for increases

Table 69--Marketwide and individual handler pools in Federal order
markets 1/

Year	Number of markets	Percent of markets	Percent of producer receipts	Percent Class 1
		Marketwide pools		
1946	22	73	79	63
1950	29	74	85	56
1955	46	73	85	59
1961	51	75	91	59
1965	65	89	93	62
1977	44	94	99	53
1981	45	94	99	46
		Individual handler pools		
1946	8	27	21	76
1950	10	26	15	73
1955	17	27	15	79
1961	14	17	9	78
1965	2/8	11	7	82
1977	3	6	1	80
1981	3	6	1	80

1/ Numbers apply to pooling arrangements in effect at the end of
the year.
2/ The only large market with individual handler pooling was
Delaware Valley (Philadelphia area), which accounted for half of all
producer receipts in individual handler pools.

Source: 1946-61: Norse et al, 1962.

in prices flooded in. Costs of production were increasing due
to the rising general price and wage level and higher prices were
necessary to bring forth increased production. During the 1941-42
fiscal year, 33 amendments were issued affecting 18 of the orders
then in effect, many of these for temporary price increases.

In the Midwest, competition from plants manufacturing butter,
cheese, and canned milk threatened to create a shortage of milk for
fluid use. It was decided to establish Class 1 prices by formula at
a fixed amount above the price of manufacturing grade milk or of
manufactured products (U.S. Agric. Mktg. Admin., 1942, pp. 65-66).
This formula method of fixing Class 1 prices was very generally
adopted for Federal orders throughout the country during and immediately after World War II (Harris and Hedges, 1948, p. 9).

In the immediate postwar years, there was a search for different
approaches in many quarters. Milk markets were still independent,

thirties. Still, long-distance movement of milk was expensive, so prices in a given market could vary quite a bit without being "out of line" with other markets. The Boston Milkshed Price Committee-- a group of economists employed by New England universities, cooper- atives, and proprietary handlers--came up with a new approach to Class 1 pricing. It was an attempt to get away from fixed prices or formulas which tied Class 1 prices to manufactured product prices, which had fluctuated wildly when price controls were removed. It set out to relate changes in factors affecting supply and demand to the Class 1 price. With only slight modification, the proposed economic formula was adopted in the Boston, Fall River, and Lowell- Lawrence markets. As other New England markets became regulated, they were linked to it.

Formulas which differed in detail but not in principle were adopted in New York, Philadelphia, and other major northeastern markets, and in New Orleans and San Antonio. (The New York formula included only the Wholesale Price Index.) Midwestern markets, closer to the manufacturing milk markets, continued to operate with formulas which related Class 1 prices to butter-powder prices or to prices paid by condenseries.

Economic formulas, except in San Antonio, included a "supply- demand adjuster," which was intended to vary the price actually paid from that provided by the formula in order to reflect local supply and demand conditions. In general, Class 1 prices were to be lowered when Class 1 utilization (seasonally adjusted) fell and were raised when it increased. This was a classic example of the tail wagging the dog and led to the comment by Ed Gaumnitz, a veteran of the milk- pricing wars, that he couldn't care less what was in the economic formula as long as the supply-demand adjuster did its job.

Supply-demand adjusters were added to most Federal orders-- either directly or by relating prices in smaller markets by formula to those in larger markets with a supply-demand adjuster. By 1960, most Federal order markets were in fact operating with supply-demand adjusters.

The function of a supply-demand adjuster as a part of an eco- nomic formula differed quite a bit from that in markets where base prices were still tied to manufacturing milk values. The effects were also larger than in other orders. In Boston and New York with market- wide pools and a consequent incentive for unregulated country plants to seek affiliation with the order, the supply-demand adjusters oper- ated frequently. In Philadelphia with its individual handler pool, it seldom operated--never from 1951 to 1959.

In the Midwest, all major markets had supply-demand adjusters by 1960 and many of the secondary markets were tied to those of major markets. Since the basic prices were tied to manufacturing milk values, the adjusters had relatively minor effects on blend prices. Changes in Class 1 utilization had greater effects as supply and demand conditions in local markets changed (Williams et al, 1962).

The experience with such adjusters indicated that, although the principles were undoubtedly correct, the difficulties of designing an appropriate mechanism limited their usefulness. Conflict inevitably arose between the needs to maintain intermarket price alignment and to adjust prices to reflect local supply-demand conditions.

Economic formulas which included supply-demand adjusters were in use in all Northeastern markets by 1955, while only a few Midwestern markets had supply-demand adjusters (with base prices still tied to manufacturing milk values). In 1960 and 1965, nearly all Midwestern markets had supply-demand adjusters. Federal order Class 1 differences were higher in the Midwest than the Northeast in 1947-55, but the economic formulas raised them higher in the Northeast than in the Midwest in 1960 and 1965 (table 87).

In the sixties, a growing awareness of the closer association of markets throughout the country—because of the improved transportation facilities, roads, and trucks—led to the present system of Class 1 prices throughout the country east of the Rockies based on upper-Midwest-plus-transportation. A basing point approximately at Eau Claire, Wisconsin, was selected and Class 1 differentials generally related to it, although there were some differences to account for local supply-demand conditions. Markets west of the Continental Divide were, in fact, separate and prices were related to supply-demand conditions in that area.

Changing Objectives. The original milk marketing agreements of 1933-34 were fairly straightforward efforts to provide relief to hard-pressed dairy farmers by raising milk prices on a market-by-market basis. While prices could be raised, providing some temporary relief, soon other problems arose. Nearly all fluid milk markets were figuratively-speaking awash in a sea of milk only too anxious to be added to that market's milk supply. Raising prices for those producers furnishing milk to a given market created much greater differences in prices between those "in" and those "out" of the market... and greater pressures to get in.

The objectives and approach were changed sharply in late 1934. The underlying philosophy has guided the development of Federal milk orders most of the time since then. The basic idea was that an unfettered milk market composed of thousands of producers some of whom were organized into one or more cooperatives confronted 50-200 processors including a handful of large firms prevented economic forces from working themselves out in the manner envisioned by the perfectly competitive economic model. The authors of this approach regarded the less-than-satisfactory conditions that had beset the dairy industry even before the Depression as being caused by the concentration among distributors and the consequent too-high returns. "Their approach was to establish such prices as would control and direct the competitive factors toward the end that they thought the freely competitive system would attain." (U.S. Agric. Adj. Admin., 1939a, pp. 200-208).

Marketing agreements had originally been thought of and operated as tripartite agreements between the government, producers, and dealers. There was something in the agreement for everyone...higher and more uniform prices to producers and more uniform prices and margins for dealers. With the withdrawal of resale price fixing, the incentives for dealers to participate were minimized. Agreements then became a joint enterprise between the Government and producers as represented by cooperatives.

This joint enterprise was recognized in the 1935 legislation and continued in the 1937 act. Positive action was required by both the Government and producers to impose or revise regulation. Administration of milk marketing orders was wholly in the hands of the Government, unlike fruit and vegetable orders. The basic procedures for milk marketing orders became--

o One or more cooperatives proposed the terms of a new market order or an amendment.
o Producers, dealers, and the public presented their views at a public hearing, arguing the merits or lack thereof of the proposed order and presenting counter-proposals.
o The Government decided what order action should be taken, in its view, and published a proposed order or amendment.
o Producers voted on whether or not to accept the Government's proposal, with cooperatives voting for their members.

The effect was that everyone interested had an opportunity to make his views known, both at the hearing and in written comments at later stages. The final order had to be acceptable to both the Government and to two-thirds of the producers as represented by their cooperative(s). The Government could always prevent regulation by not issuing an order or by withdrawing one which it felt no longer served the purposes of the Act. Cooperatives could not force acceptance of a regulation which they desired but they could prevent one they disliked by voting against it or asking that an existing order be withdrawn. The cost to a cooperative of such an action was the loss of the entire order, so there was considerable room for maneuver between the Government and the cooperative(s) over disputed provisions of an order.

From Local Markets to a National Price Structure. The localized Federal order markets of the thirties and forties grew closer together in the economic sense in the fifties. In the early sixties, a more closely coordinated system of Class 1 prices was adopted in which changes in national supply-demand conditions was reflected simultaneously in all Class 1 prices. The Minnesota-Wisconsin (M-W) price for unregulated milk for manufacturing uses became the mover of all Class 1 prices and local supply-demand adjusters were phased out. The M-W price was the average price paid for manufacturing grade milk delivered to plants in the heart of the surplus production region. Prices there were determined under competitive conditions, the major constraint being that provided by the price support program which effectively set a lower limit to prices, as intended. In times of national milk surplus, the M-W price was near the support level, usually a bit below it. When milk supplies became tighter--either seasonally or on an annual basis--buying prices rose above the support level as plants bid against each other for milk supplies.
Thus, the M-W series had much to recommend it, although it certainly wasn't perfect--it provided an indicator of changes in the overall supply-demand situation; it was competitively determined; it was a measure of the cost of alternative milk supplies from the upper

Midwest; and it provided a means of coordinating changes in Federal order class prices with changes in price levels under the dairy price support program.

A structure of Class 1 prices east of the Rockies was created by adding differentials based on cost of transportation from the upper Midwest.

As a result, the character of prices established under Federal orders became quite different. They were no longer subject to frequent changes because of altered supply-demand conditions in individual markets. Instead, they became a coordinated system of prices for all markets where the major factors moving prices up or down were changes in the national supply-demand situation and the price support level. Changes in individual order prices were made only in the context of a system of prices for all markets. There was less opportunity to change prices for an individual order or for a group of markets in response to local or regional supply-demand conditions. It also required more extensive use of national hearings for all orders.

The transformation of the Federal order system from one of loosely related local markets to a near-national system involved more than the Class 1 price structure. From the mid-sixties on, policy reflected the view of Federal order management that all milk was becoming Grade A and that the Federal order system should adapt. This policy primarily affected the upper Midwest, since that was where most of the Grade B milk was produced. Pool plant requirements were eased, making it less burdensome for a manufacturing plant to participate in the Federal order pool. Removal of individual market supply-demand adjusters reduced pressure on prices from increased receipts of milk newly converted from Grade B to Grade A. New orders were established in the upper Midwest and later merged into large regional orders. All of these actions had the effect of changing Minnesota from the State where 8 percent of all milk sold wholesale by farmers in 1960 was regulated by Federal orders to 61 percent so regulated in 1981.

Class 1 sales were 62-65 percent of producer deliveries in all Federal orders in the late fifties and sixties. The Class 1 percentage was down to 58 percent in 1975 and 46 percent in 1981. In the Upper Midwest order, 14 percent of producer milk was used in Class 1 in 1981. Far from all of the added milk came from conversion from Grade B, but that was a major part.

Defining Class 1 Products. The introduction—both potential and actual—of new or modified fluid beverage products has caused controversy in the past and no doubt will continue to do so. Each of these products presents Federal order management with problems in classification and pricing: is the item to be regarded as a Class 1 product and how are prices for it to be established? The question has arisen for—

o Concentrated milk in the early fifties.
o Filled and imitation milk in the sixties.
o Reconstituted milk in the late seventies.
o Ultra-high-temperature (UHT) milk in the eighties.

Concentrated milk was a fresh product with much of the water removed by technology that did not produce the cooked flavor of canned milk. It could be shipped at costs substantially below those of regular milk. (See Mathis, 1958.) Filled milk was made from fresh skim milk and vegetable fat, usually cocoanut oil. Imitation milk was made from vegetable fat, casein, and other ingredients. (See Butz et al, 1968.) Reconstituted milk could be made in the fluid milk plant from a variety of ingredients including cream or butter, and nonfat dry milk or condensed skim. (See Novakovic and Aplin, 1981, and Hammond et al, 1979). UHT milk is beverage milk treated by a form of pasteurization that makes it unneoessary to refrigerate the final product.

The problems of classification and pricing involve the issues of who gets what and the integrity of the classified pricing system. Producers and their organizations in the defined Federal order market have accepted an obligation to provide an adequate supply of milk for consumers in that market in return for somewhat higher prices than are paid to manufacturing milk producers. Those prices must also be adequate to cover the additional costs of serving and balancing the fluid milk market. If ingredients from other markets—including nondairy ingredients in some cases—are allowed to be used in milk-like beverage products at unregulated prices lower than regular milk, producers of those ingredients share in the benefits of the classified pricing system without sharing in the costs. This is another case of externalities or the free-rider problem.

Each of the cases has been dealt with by measures which assured that the cost of ingredients was at least as high as that reflected in order prices and often higher.

Regulation of Nonpool Milk. Another externality or free-rider problem dealt with nonpool milk—non-order milk coming into a market to skim off some of the Class 1 sales without bearing any of the costs of serving or balancing the market.

In the thirties, the biggest problem of this type involved producer-distributors, farmers who produced their own milk and distributed it. They were accorded special status under the legislation and, generally speaking, were unregulated except for reporting statistics as long as they took care of their own fluid needs and their own surplus, if any. Some very large operators with hundreds of cows in Florida and western Washington have come under this exemption. Treatment of producer-distributors is no longer much of an issue except occasionally in markets with these larger operations.

More controversy has arisen over the treatment of other-order or unregulated milk purchased by regulated fluid milk handlers. A number of devices have been used to prevent or minimize the ability to free-ride in a marketwide pool. The most common was some form of compensatory payment which usually caused such nonpool milk to be more expensive than pool milk. The specific compensatory payment provisions of the New York-New Jersey order were declared illegal by the Supreme Court in 1962 (Lehigh Valley Cooperative v. United States, 370 U.S. 76) as constituting an illegal trade barrier. The Department subsequently suspended compensatory payment provisions in other orders. The decision was regarded in some quarters as opening up Federal order markets to this form of free-riding on a large scale.

The 1964 order decisions put in place revised terms dealing with nonpool milk which have withstood legal challenge. The effects of these provisions are somewhat different than the earlier ones but they have generally discouraged free-riding.

Cooperative Role in Federal Orders. Cooperatives have always had a significant role in Federal milk marketing orders and agreements. From the beginning, the Department of Agriculture held that the initiative for a market order must come from the cooperative or cooperatives. If no request for a market order was received, conditions in the market must be satisfactory to producers. It took some time to develop an attitude on the part of cooperatives that market orders were a joint enterprise with the Government. Originally, some cooperatives felt the Act was passed for their benefit and that each cooperative had the right to use it as it saw best (U.S. Agr. Adj. Admin., 1939a, p. 207). Gradually, the approach of jointly seeking a solution to problems became more prevalent. The necessity for the cooperative to present a justification for its proposal at a public hearing helped to reinforce the joint approach.

Over 80 percent of the producers in Federal order markets have been cooperative members since at least 1960 (table 70). In the early sixties, there was some tendency for larger producers to remain outside the cooperatives, but by 1978 cooperative and non-cooperative members averaged almost the same size. The omission of New York and Chicago in the earlier years undoubtedly influenced these statistics, since they are the largest markets.

The tendency for larger markets to have more cooperatives is evident in table 71. In December 1980, the major cooperative association represented at least 80 percent of the producers in a third of the orders, but these were in small markets accounting for only 17 percent of the milk marketed under all orders. On the other hand, 13 percent of the orders in which the major cooperative represented less than 30 percent of the producers were in large markets accounting for 41 percent of the total milk marketed under orders.

State Milk Control

Three-quarters of the 50 States have essayed State regulation of producer milk prices at some time since 1933. In practically every other State with any significant dairy industry, attempts have been made to pass enabling legislation. At the close of World War II, 16 States were regulating producer milk prices. The number rose to 20 in 1965-67 and has now declined to 15 (table 72). The 16 States regulating milk prices in the decade following World War II accounted for nearly a quarter of the fluid grade milk supply. The 15 controlled 17 percent of all fluid grade milk sold to plants and dealers in 1982 (table 64).

In the early years, States regulating producer prices were concentrated in the Northeast (table 73). Several States in the Midwest controlled producer prices during the thirties, but all gave up those efforts by 1941. In the late sixties, North Dakota started

Table 70--Proportion of Federal milk order producers belonging to cooperative associations and proportion of producer deliveries under Federal milk orders marketed by cooperative members, by regional groups of orders, December 1960-80

Region 1/	Producers belonging to cooperatives					Producer deliveries marketed by cooperative members			
	1960	1965	1970	1975	1980	1960	1965	1975	1980
	--Percent--					--Percent--			
New England	83.0	91.3	94.7	86.6 ⎫	69.2	70.6	91.4	83.0 ⎫	68.6
Middle Atlantic 2/	63.4	76.8	72.5	70.4 ⎭		62.0	73.0	70.2 ⎭	
South Atlantic	81.3	87.9	99.1	94.4	91.4	85.5	86.5	94.7	93.2
East North Central 3/	86.1	87.9	88.7	91.3	89.6	85.5	86.3	91.5	90.5
West North Central	92.3	94.9	94.6	93.3	87.3	92.2	94.3	93.5	88.3
East South Central	87.5	89.5	95.3	87.6	82.8	86.7	87.0	85.0	82.1
West South Central	83.5	85.3	92.5	91.8	91.5	83.3	85.2	88.0	87.0
Mountain 4/	88.3	93.1	97.0	94.4	95.7	87.6	87.3	90.3	92.0
Pacific	72.9	72.4	72.5	81.8	88.9	70.0	69.9	77.4	85.3
All regions combined	84.4	88.3	86.5	86.4	83.6	77.6	86.7	85.9	84.0

1/ Regional total have not been adjusted over time for marketing area changes.
2/ Excludes New York-New Jersey in 1960 and 1965.
3/ Excludes Chicago in 1960.
4/ Excludes Colorado Springs in 1960.

Source: Krueger and Woldenberg, 1979, p. 37; Krueger and Currier, 1982, p. 43.

Table 71--Frequency distribution of Federal milk order markets and proportion of producer deliveries marketed under orders according to the percent of producers belonging to the largest cooperative serving the market, December 1960, 1970, and 1980

Percent of producers belonging to largest cooperative	Federal milk order markets				Federal milk order deliveries	
	Number	Percent of total	Cumulative		Percent of total 1/	Cumulative percent of total
			Number	Percent of total		
			December 1960 2/			
100	4	5	4	5	1	1
90 - 99	13	17	17	22	6	7
80 - 89	13	17	30	39	7	14
70 - 79	12	16	42	55	14	28
60 - 69	7	9	49	64	3	31
50 - 59	10	13	59	77	8	39
40 - 49	8	10	67	87	7	46
30 - 39	6	8	73	95	5	51
20 - 29	4	5	77	100	49	100
0 - 19	0	0	77	100	0	100
			December 1970			
100	9	15	9	15	3	3
90 - 99	12	19	21	34	8	11
80 - 89	11	18	32	52	11	21
70 - 79	9	15	41	66	7	29
60 - 69	4	6	45	73	9	38
50 - 59	7	11	52	84	14	52
40 - 49	2	3	54	87	12	64
30 - 39	3	5	57	92	13	77
20 - 29	5	8	62	100	23	100
0 - 19	0	0	62	100	0	100
			December 1980			
100	4	9	4	9	1	1
90 - 99	4	9	8	17	2	3
80 - 89	7	11	15	32	10	13
70 - 79	4	9	19	40	6	19
60 - 69	8	17	27	57	13	32
50 - 59	5	11	32	68	8	40
40 - 49	9	9	41	87	19	59
30 - 39	0	0	41	98	0	59
20 - 29	5	11	46	100	29	88
0 - 19	1	2	47	100	12	100

1/ Producer deliveries in those markets that comprise each frequency stratum (percent of producers belonging to largest cooperative) as a percent of total producer deliveries in all markets combined.

2/ Excludes Chicago, Colorado Springs, and New York-New Jersey.

Source: Krueger and Woldenberg, 1979, p. 38; Krueger and Currier, 1982, p. 44.

Table 72--Number of States controlling producer and resale prices for milk, 1933-1981

Year	Producer prices		Resale Prices			
	Total 1/	Exclusive 2/	Any 3/	Wholesale	Retail Any:	Stop-loss 4/
1933	8	8	6	6	6	0
1935	21	21	20	20	20	0
1940	20	20	17	17	17	0
1945	16	16	14	14	14	0
1950	16	16	12	12	12	0
1955	16	16	12	12	12	0
1960	16	16	13	13	14	0
1965	20	19	14	14	13	1
1970	18	17	15	15	14	2
1975	18	16	13	10	11	2
1981	16	14	7	5	6	1

1/ Includes States regulating concurrently with Federal orders for all or part of the State.
2/ Includes States with exclusive regulation of at least part of the State.
3/ States regulating either wholesale or retail prices or both.
4/ States establishing minimum prices well below existing levels.

Source: Black, 1935; U.S. Agric. Mktg. Serv., 1955 and 1959; Spencer and Christensen, 1954; U.S. Econ. Res. Serv., 1962a; Mathis et al, 1965; Mathis and Favel, 1968; Forker and Aplin, 1970; Mathis et al, 1972; Shaw and Levine, 1978, Krueger, 1981.

Table 73--States controlling producer prices for milk, by region, 1935-1981

Region	Number of States controlling producer prices in --									
	1935	1940	1945	1950	1955	1960	1965	1970	1975	1981
Northeast 1/	11	10	10	10	10	8	8	7	7	7
Midwest	4	3	0	0	0	0	0	1	1	1
South	2	4	3	3	4	5	7	5	5	4
West	4	3	3	3	2	3	5	5	5	4
United States	21	20	16	16	16	16	20	18	18	16

1/ Includes Maryland and Virginia.

producer price regulation in the western part of the State which is not under the Federal order. In the sixties, additional States in the South and West joined the ranks of those regulating prices of producer milk.

Classified pricing was generally used under State milk control, although not universally so. A certain amount of flat pricing was permitted in some States.

Marketwide pooling was much less generally used. The original legislation in the Northeastern States did not, in any instance, authorize it specifically. Later amendments to the legislation authorized it in New York in 1937, Rhode Island in 1939, and Massachusetts in 1941. The authority was used in New York in the Buffalo and Rochester markets, and in Massachusetts in New Bedford only. It was not used in Rhode Island (Spencer and Christensen, 1954). In 1962, only Louisiana, Mississippi, and New York used marketwide pooling and, in 1976, California, New York, Hawaii, and Vermont (U.S. Econ. Res. Serv., 1962a; Shaw and Levine, 1978).

Major markets had long carried the surplus for secondary markets and, when both Federal and State regulation came along in the thirties, it wasn't long before this situation became institutionalized. Thus, the Boston pool carried the reserves for secondary markets throughout New England, as the New England order now does for Maine and Vermont. The market pool in the New York Federal order carried the reserves for the upstate markets as it still does. In addition, New York and other adjoining Federal order pools carry the reserves for the State-regulated markets in central Pennsylvania. Until Philadelphia gave up its individual handler pool in 1970, much of its reserve was carried by the New York-New Jersey pool.

The same situation has applied for a number of years wherever, across the country, there were pockets of State-regulated milk, except in California which is a nation unto itself in milk as in other matters.

A wide variety of methods of establishing producer prices have been used by the States. A number have based their prices directly on measurements of cost of production or of changes in costs. Some-- as in northern New England--tied their prices directly to price levels of adjoining Federal orders. Economic formulas have been used by a few. Others relied on the hearing process.

The resulting price levels were frequently above those in nearby Federal order markets. Indeed, in the fifties and early sixties in markets which enjoyed the benefits of both Federal and State regulation (except for New York and New Jersey where the Federal and State orders were identical), the State minimums were frequently above Federal levels. The early examples of so-called "over-order premiums" were largely due to State regulation. Since the mid-sixties, instances of State prices higher than Federal order prices for a given market have mostly disappeared, although alignment of Federal order and State prices, in areas where both do not regulate, is less than perfect.

In many--not all--cases, State milk control was a deterrent to the growth of bargaining cooperatives. This can be seen in States as geographically diverse as Maine, the Carolinas, Georgia, Alabama, and California. Especially where States utilized individual handler pooling--or none at all, as in California with its "contract" system until the late sixties--there was little role for bargaining coopera- tives and they played little if any part in milk marketing. This feature, together with the guaranteed margins provided by resale price control, furnished strong incentives for dealer support of State milk control.

Some of the effects can be seen in table 74. In States with nearly all milk under State regulation, less than half the milk was marketed by cooperatives in 1973. Where the States played little or no part, cooperatives were much more important.

Table 74--Milk marketed by cooperatives and State milk control, 1973

Percent of fluid grade milk sold to plants and dealers under state milk control	Number of states	Percent of fluid grade milk sold to plants and dealers which is sold by cooperatives
90-100	6	46
70-80	2	54
35-55	4	74
0-10	38	87
United States	50	81

Source: Special tabulations by George C. Tucker, ACS.

Resale Price Control. In the depths of the Depression, nearly all States started out by fixing prices at all levels of the market- ing system--prices to producers, wholesale prices (to stores and restaurants), and retail prices (home delivered and out-of-store). Since World War II, the number of States controlling resale prices has declined a bit faster than the number of those setting producer prices (table 72). Currently, 7 States control resale prices. One controls only wholesale prices and two only retail prices.

Until fairly recently, resale price control consisted in essence of price fixing. Some States set minimums and some both minimums and maximums, but almost universally the minimum was the market price. In the late sixties, Vermont and New Jersey changed their policy to set only stop-loss minimum retail prices and, until recently, retail prices in those two States were usually considerably above the mini- mums.

In the seventies, several other States dropped resale price fix- ing entirely (California, Vermont, South Dakota, Wyoming, and Ala- bama) and Virginia and South Carolina stopped fixing retail prices. Only Maine, Pennsylvania, North Dakota, and Montana still fix prices at all levels.

Resale price control was originally the quid pro quo for deal-ers, as producer price control was for farmers. Where the legisla-tion provided standards, they were generally based on the costs of processing and distribution, which the control agency typically defined as either the average of all dealers or of smaller dealers. The effect was to maintain distributor's margins at higher levels than in markets where resale prices were not regulated. This effect was found by the Temporary National Economic Committee in the late thirties (Waite et al, 1941, pp. 119-159). Other studies over the years reached similar conclusions. In 1969, Hallberg found that the average marketing margin in 22 markets with fixed resale prices was 4 cents per half gallon higher than in markets without resale price fixing or prohibition of sales below cost. These margins include the effects of different proportions of store and home delivered milk and of containers of various sizes (table 75).

Table 75--Average whole milk marketing margins in 80 markets, by type of State regulation, 1969

Type of State regulation	:	Number of markets	: :	Marketing margin (cents per half gallon)
None	:	41		25.9
Sales below cost prohibited	: :	17		25.1
Resale prices fixed	: :	22		29.9

Includes store and home-delivered milk in all containers.
Source: Hallberg, 1975, p. 3.

Removing the effects of varying proportions of home delivered and store sales (by using the same percentage in all markets) and including the effects of structural and institutional factors in 144 markets in 1967-69, price setting by State agencies was the most highly significant variable influencing margins, statistically speak-ing. All other things being equal, in markets where prices had been set under State regulation for 20 years or longer, marketing margins averaged 2.4 cents per half gallon higher than in other markets (Man-chester, 1974a, pp. 27-8; see also Greco, 1982; Shaw et al, 1975; and Masson and DeBrock, 1980).

One of the major conflicts in resale price regulation since its inception has been over the existence and size of the store differen-tial--the difference between the minimum home delivered price and the minimum out-of-store price. Store differentials were typically less common and smaller in markets under resale price control (Spencer and Christensen, 1954, p. 101). In recent years, with the decline in home delivery, the question has become moot.

By fixing marketing margins, State resale price control helped to keep smaller fluid milk processors in business for a considerable time. Plant numbers dropped sharply everywhere between 1948 and 1964, but the decline was significantly more rapid where resale price fixing did not exist. But after 1964, plant exits accelerated in states with resale price fixing.

	Decline in fluid milk plants	
	1948 to 1964	1964 to 1979
	- - - - - - - Percent - - - - - - -	
States without resale price control	55	72
States with resale price control for all years	40	78
States with resale price control for 70-82 percent of the years	57	63
States with retail price control for 12-47 percent of the years	47	72
All States with some resale price control	44	72

At the same time, resale price control provided an attractive opportunity to supermarket chains to integrate into fluid milk packaging (see Chapter 6). This provided pressure on large plants, not small, so that numbers were little affected.

Restricting Competition. Inevitably, regulation of the prices at which milk was sold at wholesale and retail also involved regulation of other competitive practices. Discounts were obviously a part of the price, so they required immediate regulation. Nonmonetary services to retailers also had to be regulated, or the price regulation would be undercut. In-store equipment (milk coolers) and in-store labor were the most prominent such services, but there were many others.

There is a strong tendency for the objective of preserving competition to be translated into the preservation of competitors, which soon came to mean those now in the business. That can lead to the exclusion of new entrants. In a few States, rules for licensing of milk dealers became restrictive. New York's milk control law provided that additional licenses should not be issued when they might lead to "destructive competition." This provision was interpreted very tightly by the New York authorities, in one case leading to denial of a license in Geneva where only two dealers were already licensed. Since licenses were issued for particular markets, when markets began to grow in size in the fifties, a dealer needed a new license to expand the scope of his distribution. These too were hard to get. (Spencer and Christensen, 1955, pp. 38-41, 90-92.)

In a few other States, somewhat similar licensing regulations were used to restrict the entrance of new firms. The Virginia situation parallels that in New York fairly closely. In Oregon, restrictive licensing was overturned by the Oregon Supreme Court in 1951, in a case involving the expansion of Safeway Stores distribution from its Portland plant into Salem.

Georgia milk control attempted to limit competition through limits on the issuance of licenses and other, more imaginative means. In one case, a price-cutting firm in Savannah was disciplined by removal of price controls from the Savannah market for a short period, while the larger competing firms carried on a price war. Only 17 days were required to exhaust the resources of the price-cutter (Harris, 1966, pp. 58-60). State controls quickly returned to the Savannah market.

Many states expanded the regulation of resale prices into regulation of trade practices. Others embarked on trade practice regulation alone. All of this was a response to the competitive pressures discussed in Chapter 7. Such laws were on the books in 28 states in 1976 (Shaw and Levine, 1978). Probably they are effective in only a handful where enforcement is vigorous. Eight require price filing, which tends to maintain prices more effectively than other means. Nearly all prohibit sales below cost and eight define cost by a percentage markup. Sometimes the markup has been set at a high level. The effects of such regulation are likely minimal in most of the country, but there are some exceptions.

PRICE SUPPORTS

The dairy price support program has undergirded the entire price structure for milk sold by farmers to processors since early in World War II. The basic program has been unchanged since that time, although there have been many modifications of details. The support prices are achieved through Commodity Credit Corporation offers to buy butter, nonfat dry milk, and cheese at prices designed to return the support price to manufacturing grade milk producers on average. Thus, the price support program has directly provided a floor under the price of milk used to manufacture these products and, indirectly, supported the price of all milk. The tie has been even closer since the late sixties when Class 1 prices in all markets were tied directly to manufacturing milk values.

Price supports for manufactured dairy products first went into effect April 3, 1941, when the Department of Agriculture announced that it would support prices of dairy products through June 30, 1943, by open market purchases of butter in Chicago for 31 cents per pound (Rojko, 1957, p.156; U.S. Bu. Agric. Econ., 1943). Under the Steagall Amendment of July 1, 1941, price supports at not less than 85 percent of parity became mandatory for all nonbasic commodities (the basic commodities were cotton, tobacco, and the major grains) for which the Secretary of Agriculture requested by public announcement an increase in production to meet wartime needs. The required support level was raised to 90 percent of parity in October 1942 and supports extended for two years beyond the duration of hostilities. Price support became mandatory for manufactured dairy products on August 29, 1941, when support prices were announced for evaporated milk, nonfat dry milk, and cheese, and for butter on November 28, 1942. These supports were required until December 31, 1948.

During the war years, price supports had only a psychological effect on the market. Prices at which USDA and the armed forces bought products were well above price support levels. Butter purchases under Lend Lease authority began March 28, 1942.

In order to hold down prices to consumers during the war, subsidies were paid directly to farmers producing milk and butterfat, to manufacturers of butter and cheddar cheese, and to milk handlers in areas with a shortage (Rojko, 1957, pp. 156-157).

After World War II, price supports at 90 percent of parity were extended through 1949 by the Agricultural Act of 1948. The 1949 Act established the permanent program which still continues. It directed the Secretary to set a support price within the range of 75 to 90 percent of parity which would bring forth an adequate supply of milk to meet the needs of American consumers (Emery et al, 1969). In 1973, the objective was broadened to include the preservation of sufficient productive capacity to meet anticipated future needs.

The minimum support price for milk has been raised from the 75 percent level in the basic legislation to 80 percent by Congress four times (table 76). In September 1960, the minimum support price was raised 16 cents per hundredweight to $3.22, which was 80 percent of parity as of April 1, 1960. In August 1973, it was raised to 80 percent of parity through March 1975. The minimum was again raised to 80 percent in September 1977 to be effective through March 1979. In November 1979, the minimum was again raised to 80 percent for 2 years. In 1957, 1975, and 1976 legislation raising the minimum level was passed and vetoed.

The 1981 Act, passed at a time of large surpluses, utilized a set of triggers relating the minimum support level to the size of CCC purchases. As long as large removals continued, the support prices were specified in dollar terms with the 1981-82 price at the 1980-81 level of $13.10 per hundredweight which was 72.9 percent of parity in September 1981 and minimal increases thereafter. Only if surpluses declined to stated levels would supports at 70 or 75 percent of parity be required. With continued surpluses, 1982 legislation froze prices for two years and imposed fees on milk producers to partially offset Government costs. Part of the fee was tied to a supply control plan.

Support prices were set annually at the beginning of the marketing year and were effective throughout the marketing year, unless the Secretary chose to raise them. The Food and Agriculture Act of 1977 required adjustments in the support price to reflect any change in the parity index during the first 6 months of each marketing year. This had the effect of raising the support price in the middle of the marketing year to reflect increases in the index of prices paid by farmers. The last such increase, which would have been effective April 1, 1981, was eliminated by Congressional action.

Until 1970, the Secretary was required to support both the price of milk and the price of butterfat in farm-separated cream. Since the butterfat support price effectively set the floor under the value of butterfat in all products, the legal minimum support level constrained not only the overall milk price support level but also the relative values of butterfat and solids-non-fat. Price supports were not always set at the same percentage of parity for whole milk and for butterfat in farm-separated cream. During the Korean War, price supports for butterfat were at a higher percentage of parity than

Table 76--Manufacturing milk: Comparisons of announced support prices and U.S. average market prices paid to producers, marketing years

Marketing year beginning in-- 1/	Date effective 2/	Support level			Average market level		
		Percentage of parity equivalent		Price per 100 pounds	Price per 100 pounds	As a percentage of parity equivalent	
		Minimum	Announced 3/			In month prior to marketing year	Average during marketing year
		Percent	Percent	Dollars	Dollars	Percent	Percent
1949		90	90	3.14	3.14	90	89
1950 4/		75	81	3.07	3.35	88	85
1951		75	86	3.60	3.97	94	93
1952		75	90	3.85	4.00	93	95
1953		75	89	3.74	3.46	83	84
1954		75	75	3.15	3.15	75	80
1955		75	80	3.15	3.19	81	82
1956		75	82	3.15	3.31	86	84
	4/18/56	75	84	3.25			
1957		75	82	3.25	3.28	83	82
1958		75	75	3.06	3.16	77	77
1959		75	77	3.06	3.22	81	81
1960		75	76	3.06	3.31	83	83
	9/17/60	80	80	3.22			
	3/10/61	80	85	3.40			
1961		80	83	3.40	3.38	83	82
1962 5/		75	75	3.11	3.19	76	76
1963		75	75	3.14	3.24	77	77
1964		75	75	3.15	3.30	77	78
1965		75	75	3.24	3.45	80	79
1966		75	78	3.50	4.11	92	90
	6/30/66	75	89.5	4.00			
1967		75	87	4.00	4.07	88	87
1968		75	89.4	4.28	4.30	90	87
1969		75	83	4.28	4.55	88	86
1970		75	85	4.66	4.76	87	85
1971		75	85	4.93	4.91	85	82

Table 76--Continued

Marketing year beginning in-- [1]	Date effective [2]	Support level			Average market level		
		Percentage of parity equivalent		Price per 100 pounds	Price per 100 pounds	As a percentage of parity equivalent	
		Minimum	Announced [3]			In month prior to marketing year	Average during marketing year
1972		75	79	4.93	5.22	84	80
1973	3/15/73	75	75	5.29	6.95	99	91
	8/10/73	80	80	5.61			
1974		80	81	6.57	6.87	85	78
	1/ 4/75		89	7.24			
1975		75	79	7.24	8.12	89	84
	10/ 2/75	75	84	7.71			
1976		75	80	8.13	8.52	84	82
	10/ 1/76	75	81	8.26		80	80
1977 [6]		75	82	9.00			
1977		80	82	9.00	7/8.77	85	79
1978	4/ 1/78	80	[8]80	9.43			
		80	80	9.87	9.30		
1979	4/ 1/79	80	[8]78	10.76	10.86	88	80
		75	80	11.49			
	11/28/79	80	80	11.49			
	4/1/80	80	[8]79	12.36	11.75	82	76
1980		80	80	13.10	12.72	78	73
		75	75	13.49			
1981	10/21/81	--	72.9	13.10			

[1] Start of marketing year April 1, 1951-77, October 1, 1977 to present. [2] If other than start of year. [3] The actual percentage of the parity equivalent price published in the month before the marketing year. In some cases the announced percentages, based on forward estimates of parity, were slightly different. [4] January 1, 1950 - March 31, 1951. [5] Beginning November 1962, parity equivalent is based on prices for all manufacturing grade milk instead of the "3-product" price for American cheese, evaporated milk and the butter-nonfat dry milk combination used before. [6] April-September transition period. [7] Adjusted to annual average fat test. [8] Semi-annual adjustment required by 1977 Act; announced support level as percent of parity in March.

those for whole milk. Thereafter, butterfat was qenerally supported at the lowest level legally possible.

During 1942-52, supports were above 80 percent of parity in every year and at 90 percent in most years. The pressure of demand from two wars kept prices above support levels in most years and removals were minimal (table 77).

Table 77--Support levels and CCC removals 1/

CCC removals as percent of marketings (milk equivalent):	1942-52 : Support : level : 80-90 : percent	1953-73 Support level		1974-78 : Suppprt : level : 80-90 : percent	1979-80 : Support : level : 70-79 : percent
		75-79 : percent	80-90 : percent		
		Number of years			
0-1.9 percent :	9	1	-	2	-
2-3.9 :	2	4	1	3	-
4-5.9 :	-	1	6	-	-
6-7.9 :	-	3	3	-	1
8-9.9 :	-	-	1	-	-
10-10.9 :	-	-	1	-	1
Total :	11	9	12	5	2
	CCC removals as percent of marketings				
Average CCC removals :	0.6	4.2	5.8	2.0	8.3

1/ Support level as percent of parity, average of current and pre-ceding marketing years.

Between 1953 and 1973, the minimum support level was 75 percent of parity, except in 1960-62 when it was 80 percent. In these 21 years, supports were set at the legal minimum in 7 years, above the minimum in 13 years, and in the 1960-61 marketing year at the minimum level part of the year and above it for the rest of the year. This was a period of secular decline in demand for dairy products and of fairly stable feed prices as a result of the feed grain price support program and excess capacity in feed grain production. During this period, when supports were between 75 and 79 percent of parity (2-year average), CCC removals averaged 4.2 percent of marketings. In the 12 years when supports were at 80 percent of parity or higher (2-year average), removals averaged 5.8 percent. While there were many differences between sub-periods, it appears that the longer-run supply-demand equilibrium level of milk prices was in the range of 75-79 percent of parity from 1953 to 1973, but lower since then.

The relatively stable feed prices provided by the price support programs ended with the Russian grain sale of 1972 and, since that time, supply-demand relationships in the dairy industry have been considerably different. Supports were at 80 percent of parity or above until 1980 and CCC removals were below 3 percent of marketings except in 1977 and 1980. Changes on both the supply and demand side contributed to this situation. The slowdown in the long-time decline in commercial demand for dairy products (see Chapter 7) was a contributor.

From the end of the Korean war through the sixties, removals of nonfat solids were consistently greater than those of milkfat (figure 5). The repeal of the requirement for support of butterfat in farm-separated cream provided greater flexibility in adjusting relative prices of milkfat and solids-not-fat. In the seventies, relative prices of the components were adjusted from time to time and removals of milkfat were higher than those of nonfat solids at some times and lower at others.

The price support program has kept prices more stable than they otherwise would have been, principally on the low side of the price cycle. Producers and resources were kept in dairy production and market prices were prevented from dropping to a very low level. Price supports have also been used to raise prices significantly above minimum levels. Sometimes such an action has merely confirmed price levels already established by supply and demand conditions, allowing policymakers to bask in the glow of higher prices created by market conditions, especially those outside the dairy industry—e.g., in feed grains or beef. Perhaps the greatest complications have arisen when price levels such as these have been continued, disregarding changed conditions in the feed grain or beef markets. The effects on the high side were more modest—market prices were prevented from rising even more than they did whenever changed market conditions made it possible for the Government to sell dairy products back to industry at prices usually 10 percent above purchase prices. In 1950 and 1951, 140 million pounds of butter purchased in 1949 were sold back to the trade. Other CCC sales in significant quantities were made in 1964, 1965, 1972, and 1975.

The program has been carried out through purchases of butter, nonfat dry milk, and American cheese. These products are widely produced and take two-thirds of the milk used in manufactured dairy products. CCC stands ready to buy these products in bulk—butter in 60 to 68-pound containers, nonfat dry milk in 50-pound bags, and cheese in 40-pound blocks or at times 500-pound barrels—at prices designed to result in a U.S. average price for manufacturing milk equal to the support price. The objective is to support only the average price, not the price to each producer. The prices received by individual producers depend upon many factors other than the support level, including plant location, product manufactured, quantity of milk delivered, the local competitive situation, and operating efficiency of the plant.

To attain the desired level of prices for manufacturing milk when product prices are at the support level, CCC purchase prices

Figure 5
Milk Solids Removed From the Market by CCC Programs*

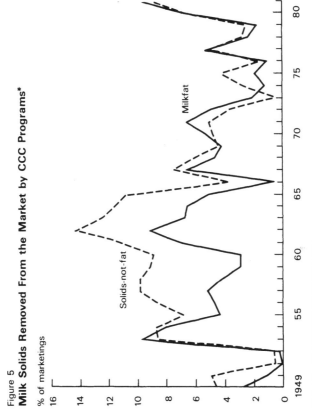

*Deliveries after domestic unrestricted sales

include "make allowances" or margins for the cost of processing milk
into butter/nonfat dry milk or cheese. These manufacturing margins
are adjusted from time to time to reflect changes in manufacturing
costs, although there is no established schedule or procedure like
that used by California. At times, as in 1981-82, there is a considerable lag in making the adjustment.

In order for such a purchase program to operate effectively,
adequate outlets must be available for the products acquired. Otherwise, excessive CCC stocks would accumulate. In the fifties and
sixties, domestic donations of food and somewhat similar foreign
donation programs provided outlets for most of the products acquired.
With food stamps replacing food distribution in the seventies, school
lunch became the only remaining sizable domestic outlet until 1981-2
when direct distribution was resumed on a limited scale. Donations
for foreign feeding programs became severely limited by budgets for
foreign aid, although CCC can donate abroad using its own funds.
Subsidized export (PIK) was a major outlet for butter and nonfat dry
milk in the sixties but this program no longer exists and Government
policy takes a dim view of export subsidies.

CONSUMPTION PROGRAMS

The Federal Government initiated a number of programs to encourage consumption of dairy products during the depression. The 1935
Act authorized purchases for distribution to schools, families, and
institutions. The National School Lunch Act of 1946 put that program
on a permanent basis, providing a subsidy for school lunches that
included milk. Free or reduced-price lunches were provided for those
who could not afford the full price. Nearly half are now provided
free or at a lower price.

School milk service was extended to non-lunch hours in 1954.
Under the Special Milk Program, low-priced milk was made available
to all students once and sometimes twice a day. A School Breakfast
Program, designed primarily for students from low-income families,
was initiated in January 1967. Significant quantities of milk were
utilized in these child nutrition programs, with much of it a net
addition to milk consumption.

The Agricultural Act of 1954 authorized CCC to subsidize purchases by the military and the Veterans Administration of dairy products acquired under price supports for use in addition to their
normal market purchases. All of these consumption programs operated
to increase modestly the demand for fluid milk and, to a somewhat
lesser extent, for other dairy products.

IMPORT REGULATION

The price support program maintains prices of dairy products
above world market levels. The prices of manufactured dairy products
in international trade—primarily butter, cheese, and nonfat dry
milk—have been low compared to the domestic prices in the producing
countries ever since the dairy industries of Europe recovered from
World War II. Each of the major European dairy producing countries
subsidizes its domestic industry and the European Community promotes

exports with subsidies of $2-3 billion per year. In these circumstances, import controls have been necessary to prevent flooding the U.S. market with foreign dairy products. (For estimates of the effects of increasing or removing import quotas, see U.S. Econ. Res. Serv., 1975; Salathe et al, 1977; Novakovic and Thompson, 1977.)

Import quotas were first authorized under Section 22 of the Agricultural Adjustment Act of 1933, which was added to the law in 1935. They did not become applicable to dairy products until later and were first imposed in 1951 under emergency legislation. Countervailing duties may also be imposed by the Secretary of the Treasury when bounties or grants are being paid, directly or indirectly, upon production or export of a product. Countervailing duties have been applied sparingly, to say the least. Effectively, they cannot now be applied to items under quota.

Under these regulations, imports of dairy products were held to modest levels in most years. In 1960, imports were 0.5 percent of U.S. production, rising to 0.75 percent in 1965. There was a sharp jump in 1966 and 1967 with greatly increased imports of butterfat mixtures. Imports in these years were 2.3-2.5 percent of production. The definition which permitted importing butterfat-sugar mixtures as "ice cream," a nonquota product, was changed and imports then held at about 1.5 percent of U.S. production until 1972. Substantially increased imports were permitted in 1973 and 1974, in response to sharply increasing milk prices in a period of rapid general inflation. These import actions helped to hold down dairy product prices, but were badly timed and contributed to wide fluctuations in milk prices in those years. Since then, imports have returned to the previous level of 1.5-1.6 percent of U.S. production.

Quotas are applied only to products that are considered to compete directly with U.S.-produced dairy products. Nearly half of the quantity and two-thirds of the value of imports was not under quota in the late seventies. In 1980 and 1981, quotas were increased by the Tokyo Round of trade negotiations and non-quota imports fell to 12-14 percent of total imports. Total imports were about the same in 1981 as in 1978 and 1979. Imports of butter, nonfat dry milk, and American-type cheese compete very directly with those products made in the U.S. and displace them essentially pound-for-pound. More exotic cheeses which are not made in the U.S. compete less directly with domestic products and are not subject to quota. Casein is a milk product not under quota. In recent years, it has entered many food markets, whereas in earlier years it was primarily an ingredient of nonfood industrial products. (See U.S. Econ. and Stat. Serv., 1981, and Chapter 12.)

U.S. quotas on imports of dairy products have become a symbol of the first magnitude in international trade negotiations. Repeated references to them in horrified tones by Europeans who are often almost giving away butter and other products in order to dispose of the surpluses engendered by their own price supports and by New Zealanders and Aussies whose subsidies are relatively small have made U.S. quotas a very touchy subject. In effect, no change is possible, any tightening (as for casein) being ruled out by trade negotiation posture and any loosening being expensive in terms of price supports.

ANTITRUST POLICY

Federal antitrust activity has waxed and waned over the years since the passage of the Sherman Act in 1890. These cycles in antitrust activity reflected changing attitudes in the Executive and Legislative Branches as to the possibility of bringing about desirable economic change through administration of the antitrust laws. While most antitrust practitioners and proponents can safely be assumed to prefer an economy characterized by small or at most middle-sized firms to one where large firms predominate, 90 years of antitrust enforcement can hardly be said to have brought about such a condition. Despite a failure to halt the growth of large firms, antitrust activity has definitely modified the behavior of American business firms over the years—certainly as compared with the behavior with firms in countries without such antitrust activity. Overt cooperative activity among business firms ("collusion" in antitrust language) especially in matters relating to price, is much less common than elsewhere in the world.

The antitrust laws existing from 1915 to 1950 were largely ineffective in preventing the growth of large firms through merger and acquisition. The Celler-Kefauver Act of 1950 plugged this gap by making acquisition of assets potentially illegal under the antitrust laws as acquisition by purchase of capital stock had been previously. The Federal Trade Commission quickly brought actions against each of the major dairy companies to halt the merger movement in fluid milk and ice cream.

FTC complaints against each of the four largest dairy companies were issued in 1956. Final decisions were reached in 1962-65. FTC decided the cases against Foremost and Beatrice and entered into consent agreements with National Dairy Products and Borden. Each of these decisions and agreements provided for a 10-year moratorium on acquisitions of fluid milk and ice cream companies without prior approval of FTC. By adopting merger guidelines for the entire industry, FTC made prior approval for acquisition a requirement for all large companies.

The apparent objectives of this merger policy for the fluid milk and ice cream industries were: (1) To stop growth through merger by the largest firms; (2) to build up the "second tier" of dairy companies; and (3) to create conditions which would persuade the firms prohibited from merging to enter new markets with new capacity—the "potential competition" doctrine.

FTC actions were effective in stopping mergers by the national dairy firms within fluid milk and ice cream processing. Merger activity declined immediately after the complaints were filed and essentially ceased after 1965. The merger policy of the FTC and the Antitrust Division constituted a per se prohibition against horizontal and vertical mergers by all large firms in the dairy industry as elsewhere. The growth strategy of large firms, including dairy firms, shifted to diversification and created the conglomerate merger movement. The largest national firms have voluntarily withdrawn from a number of fluid milk markets in response to the competitive pressures described in Chapter 6.

FTC was somewhat ambiguous about what constituted the "second tier" which was to be encouraged. The Southland Corporation, a convenience store and dairy processing firm, was permitted to grow from a Texas company to one which operated coast-to-coast, while Dean Foods Company—a much smaller regional milk and ice cream organization—was prohibited from acquiring the Bowman Dairy Company which was busy going broke in Chicago but which the FTC did not regard as meeting the "failing company" test.

The objective of creating conditions which would persuade the firms prohibited from merging to enter new markets with new capacity was an utter failure. No national firm entered a single new market by constructing new capacity. In a shrinking market, this was hardly to be expected.

In addition, the market for viable, closely-held firms was in some cases destroyed and at best severely restricted. In the early fifties, there was a brisk market for viable, closely-held, local fluid milk firms whose owners wished to sell either for personal or business reasons. To a very substantial extent, the potential buyers were national and regional dairy companies. Since then, their role as potential buyers has been much restricted and the market value of such local firms has declined drastically. In many cases, no buyer at all could be found. One result has been that, in many cases, the national or regional firm has found it possible to pick up the business of a local firm (the accounts—not the assets) at no cost rather than having to pay at least a part of the value of the firm as was previously the case. 1/

The Robinson-Patman Act prohibits price discrimination as between buyers. It provides considerable restraint particularly when dealing with large buyers. Differences in price must be cost-justified and the allowable basis for cost-justification is quite limited. Its use by FTC in a few milk cases where large processors were dealing with chainstores and particularly the threat of such use inhibits processors in such negotiations. It may well have been a contributing factor in the decision by a number of chains to build their own milk plants.

Cooperatives

Federal policy encourages marketing cooperatives through credit, technical assistance, and the operation of marketing orders. The basic legislation is the Capper-Volstead Act of 1922 which provides a partial immunity from antitrust laws for marketing cooperatives but sets limits on the extent to which they may enhance prices. The Antitrust Division has long regarded large cooperatives as anticompetitive and indeed its activities were part of the motivation for the passage of the Capper-Volstead Act. The Federal Trade Commission is a more recent convert to this view.

1/ For other views of FTC merger policy, see Mueller et al, 1976, and Kilmer and Hahn, 1978.

Growth of the regional dairy marketing cooperatives in the late sixties soon brought them to the attention of the Justice Department. Suits were filed against the three large regional cooperatives. Two of them were settled by consent decree. The third was lost by the Justice Department in District Court on most charges but reversed by the Circuit Court and is still in litigation. In 1976, a complaint was filed with the Department of Agriculture charging that any over-order payment in excess of 50 cents per hundredweight represented undue price enhancement under Section 2 of the Capper-Volstead Act. The Capper-Volstead Committee of the Department, which was charged with enforcement of the Section 2, found no undue price enhancement during the period covered by the complaint (Capper-Volstead Committee, 1976).

The staffs of the antitrust agencies took up the cudgels against the perceived evils of large cooperatives which might have some market power (U.S. Dept. of Justice, 1977; Ippolito and Masson, 1978; Masson and Eisenstat, 1978 and 1980; Lipson and Batterton, 1975). In their view, marketing cooperatives--especially dairy marketing cooperatives--had altogether too much market power. The remedy for this perceived evil was to treat them like all other corporations. The partial antitrust immunity provided by Section 1 of the Capper-Volstead Act would be modified to provide only that the formation of an agricultural cooperative through the voluntary association of its farmer members would be legal and nothing more. All subsequent activities of the cooperative would be subject to the same antitrust laws as any other corporation. This would specifically repeal the provisions that cooperatives could have marketing agencies in common, that they could merge with immunity from the antitrust laws, and that they could exchange information about prices. This is the extreme case of the "small is beautiful" doctrine, which holds that cooperatives are only acceptable if small and powerless. It would apply the existing per se prohibition against mergers by noncooperative businesses to cooperatives. Since cooperatives by their nature do not have the alternative of conglomerate growth, it would effectively prevent most growth.

IN SUMMARY

Public programs evolved as the dairy economy changed in the past 40 years. The price support program was started during World War II and made permanent in 1949. It was designed to place a floor under the price of milk by purchasing butter, powder, and cheese offered to CCC at stated prices. Both legislative and executive branches used price supports to bolster the incomes of dairy farmers, at times stimulating substantial surpluses and large CCC stocks.

Federal orders expanded and State regulation contracted until about 80 percent of fluid grade milk was under Federal regulation. The main exception was in California where State controls on producer prices are still in effect. The Federal order system changed from one of loosely interrelated local markets to a near-national system of closely related markets where order prices in all markets move together, in concert with the prices of manufacturing grade milk.

The barriers to movement of milk erected by sanitary and other regulation were largely dismantled.

Antitrust policy helped to end the merger wave in fluid milk and ice cream and to turn large dairy companies to diversification. Attempts to apply the existing prohibitions against large firms to cooperatives were somewhat less successful.

Part 4

Evaluation and Prospect

11
Changes in Market Organization and Power

Most of the components of change in market organization and power have already been discussed. It remains only to pull together the pieces. The analysis of market power deals chiefly with large corporations and large cooperatives. A major means of growth of companies in the dairy industry has been merger or acquisition, at least for the last 60 years. With adequate or more-than-adequate capacity through the period, additional capacity and volume was usually available at lower cost by acquisition than by building new capacity and competing for sales. Acquisitions by corporations were at an all-time peak in the late twenties when the National Dairy Products Corporation and the Borden Company started their growth (table 78). After dropping off during the Depression, acquisitions of more than 1,000 companies were recorded during World War II, a level never again reached. The FTC brought a virtual halt to acquisition by the eight largest dairy companies in the middle fifties.

In the sixties, firm growth and acquisition activities of the large dairy companies turned to diversification. Foremost merged with McKesson-Robbins and Hawthorn-Mellody with National Industries. The other large companies retained their corporate identities but most of their growth was outside of the dairy industry (table 29). Fairmont and Arden-Mayfair are now nearly out of fluid milk. More recently, Kraft merged with Dart.

Much of the market power of manufacturers of consumer goods comes from product differentiation through brand preference. The creation of strong brand preference has never been easy for many dairy products. It is extremely helpful in creating brand preferences to have real differences in flavor, texture, or quality as a starting point. Standardized products present a greater challenge to the company endeavoring to establish a strong brand position. The basic dairy products were standardized at an early date, the composition of butter by Federal law in 1915. The basis for branding then becomes uniformity of quality. Variations in butterfat content above the minimum were of considerable importance for whole milk in earlier times, but in the last 20 years they have virtually disappeared. Somewhat more variation is possible in butterfat content and other constituents for ice cream. Cheese flavor likewise can be varied by manufacturing techniques and aging. Thus, for most dairy

Table 78--Acquisitions of dairy companies by various groups of firms, 1921-70

Period		Number of dairy company acquisitions by--			
	:	The 8 largest : dairy : companies 1/	Other : corporations	Cooperatives	Total
1921-25	:	74	597	92	763
1926-30	:	652	1,172	127	1,951
1931-35	:	141	292	78	511
1936-40	:	319	389	63	771
1941-45	:	363	507	146	1,016
1946-50	:	243	445	167	855
1951-55	:	349	402	162	913
1956-60	:	150	363	144	657
1961-65	:	30	303	152	485
1966-70	:	17	180	120	317

1/ Borden, Kraft, Beatrice, Foremost, Carnation, Fairmont, Pet, and Arden-Mayfair.

Source: Parker, 1973, pp. 13-14.

products the basis for product differentiation is relatively weak. Brands are important for processed cheeses and higher-priced ice cream and, to some extent, for butter.

Market power for other dairy products rests on other sources. In the heyday of home delivery in the thirties and early forties, success depended more on service than on brand. As fluid milk distribution switched from home delivery to store sales, service to the store operator became paramount. Centralized purchasing by chain-stores and other supermarket groups put the emphasis on price and market power shifted from the processor to the retailer. Construction or purchase of fluid milk bottling plants by supermarket operators was the final step. The shift in ice cream distribution from drug stores to supermarkets reflected many of the same developments.

Two major developments in fluid milk markets drastically changed the position of processors. Essentially every chain of any size that had not already done so has now embarked on a central milk program which includes private label. Instead of handling processor label milk--often from three or four processors--the typical retail food chain now carries its own brand of milk and a small amount of other processor brands. This means that medium-sized processors are now effectively foreclosed from supermarket outlets except in isolated cases.

The next step was for the retail food chains to build their own milk plants. The number of such plants is still small and the share of all milk accounted for by them is still below 20 percent. However, in a period when there was no growth in the total fluid milk market these new integrated fluid milk plants added about 10 percent to the total capacity of the fluid milk processing industry. This volume was subtracted from that of the large handlers, primarily national and regional firms. This had tremendous repercussions throughout the processing business, as firms which lost a substantial share of their sales (from 20 to 60 percent in various cases) sought other outlets. It almost certainly squeezed operating margins of some processors below the long-run breakeven point for a time. At the same time, retail store margins appear to have increased.

The other major development which further squeezed processors has been the growth of large regional milk marketing cooperatives. Whatever advantages individual firms previously had on the procurement side have been drastically reduced.

In much of the dairy industry, processors have lost market power to retailers on the one hand and cooperatives on the other. The decline in numbers and growth in size of dairy marketing cooperatives is seen by many as clear evidence of their growth in market power. While the growth in size and decrease in number of dairy cooperatives has been impressive, the growth in power is a much different matter, since much of the growth in cooperative size was merely a reflection of the increase in size of milk markets due to technological and economic change.

COOPERATIVE FUNCTIONS AND POWER

The bases of potential cooperative power include:

o Control of supply, either of the volume produced or of the disposition of the supply once produced (these are qualitatively different).
o Control of the sources of milk supply, preventing access to alternative sources.
o Efficiencies derived from economies of size in manufacture of products and, more important, in managing the routing and utilization of milk.
o Efficiencies in marketing and merchandising products.
o Product differentiation, including product development.

The decline in the number of dairy marketing cooperatives has not been exceptional. Between 1954 and 1972, the number of dairy marketing cooperatives decreased by 62 percent, while the number of companies manufacturing dairy products (including fluid milk) as reported by the Census decreased 65 percent and the number of fluid milk processors, 68 percent. The increasing size of fluid milk markets and of procurement areas for dairy manufacturing plants were major contributors to the decline in numbers of plants and companies.

Dairy cooperative numbers varied between 2,300 and 2,400 during the late twenties and the thirties (table 79). They started to decline during World War II and have been declining ever since. All dairy marketing cooperatives--bargaining as well as processing-- handled 45 percent of the milk and cream sold by farmers to plants and dealers in 1936, about the same as 10 years earlier. After the

Table 79--Dairy marketing cooperatives, 1929-1980

Year	Associations	Membership	Gross business	Net business 1/
		Thousands	- - Million dollars - -	
1929-30	2,458	650		680
1935-36	2,270	720		520
1940-41	2,374	650	693	
1945-46	2,210	739	1,428	
1950-51	1,928	814	2,299	1,934
1955-56	1,762	800	3,029	2,539
1960-61	1,500	630	3,890	3,240
1965-66	1,226	507	4,866	3,833
1970-71	824	351	6,399	5,442
1975-76	556	206	9,577	8,480
1980	487	167		13,666

1/ Excluding intercooperative business.

Source: Agricultural Cooperative Service.

war the cooperatives' share began a fairly steady climb to 53 percent of the total in 1957, 65 percent in 1964, and 75 percent in 1973.

By 1960, cooperatives had generally recognized the need for centralized management of milk supplies and disposition of surplus milk. For years, bargaining associations had struggled with ways of gaining control of their milk supplies to strengthen their marketing position. As markets grew larger and the number of buyers smaller, cooperatives increasingly found themselves potentially in toe-to-toe competition with other cooperatives and increasingly vulnerable to undercutting on prices, service charges, delivery requirements, and other matters. The Lehigh Valley decision seemed to open up Federal order markets to unrestricted competition from all points. (Cook et al, 1978, p. 54.)

Reduced market protection and a period of competitive intermarket movements of milk emphasized the need for increased coordination between cooperatives in multimarket areas. While maintaining their separate identities, cooperatives in the central part of the country marketing about a quarter of U.S. milk began to form federations in the early sixties in an attempt to raise producer incomes through higher prices and to realize cost savings from better-organized movement of milk (Knutson, 1971; Tucker, 1972).

The large federated organizations served member cooperatives as a marketing agency in common, improved price alignment among markets, presented a united position at Federal order hearings, operated a standby pool for reserve milk supplies, expanded promotion of dairy products, and more effectively presented their views to legislative and executive branches.

By the mid-sixties, Federal milk orders began to reflect the increased need for more stable price alignment among markets. Individual market supply-demand adjusters were dropped. The Minnesota-Wisconsin (M-W) price for manufacturing grade milk became the price for manufacturing use in all Federal orders and then the basic formula price used in determining the Class 1 prices in all orders. Federated cooperative structures hampered bargaining efforts and could not deal adequately with problems of operational efficiency, equity among producers, and greater market stability.

By 1970, many of the member cooperatives of the two major federated organizations had merged into four regional centralized full-service cooperatives: Associated Milk Producers, Inc., Mid-America Dairymen, Inc., Dairymen, Inc., and Milk, Inc. (now Milk Marketing, Inc.).

Beginning in the late sixties, a number of small manufacturing cooperatives in Minnesota and Wisconsin joined the large regional cooperatives in that area. Among other things, members of the manufacturing cooperatives sought assured grade A milk markets, outlets to the growing cheese market, benefits from increased plant operating efficiency, and revolving of equity investments in obsolete facilities.

The organization of production and marketing of manufactured dairy products is very different from fluid milk products and there are major differences within the industry. Butter and nonfat dry milk have long been dominated by cooperatives which now produce three-quarters of all butter and 90 percent of dry milk products. Natural cheese production--long a bastion of relatively small proprietary firms--has seen a major growth by cooperatives which have increased their share to about one half.

Greatly increased diversity in cheese markets has increased the opportunities for cooperatives and other cheese manufacturers. While the processed cheese and cheese food market for supermarket distribution remains largely in the hands of two firms, the development of a widespread market for many varieties of cheese through wine and cheese shops and other similar outlets has opened many opportunities for other manufacturers. At the same time, the dramatic growth in sales of pizza and other pasta foods has expanded the market for Italian cheeses, primarily Mozzarella. The large cheese merchandisers have withdrawn from natural cheese production to some degree, conceding an increasing share to cooperatives.

After the regional cooperatives were formed in the late sixties, they consolidated manufacturing facilities, especially milk drying and cheese plants, to improve operational efficiency. In the West North Central Region, many small plants were closed and replaced by a few large, efficient plants. In response to market price fluctuations for cheese, the cooperatives converted butter-powder plants to cheese and expanded some existing cheese plants. As a result, regional cooperatives have a relatively large portion of the larger manufacturing plants.

The new capacity adds flexibility to plant operations. An estimated 15 percent of cheese capacity could be converted back to butter-powder production with minimum loss of time and expense. This flexibility increases the probability that products will continue to remain in closer price alignment then in the mid-seventies, when the value of milk going into cheese exceeded its value for butter-powder. However, this flexibility of operations is concentrated primarily in the regional cooperative plants. The national cheese corporations, single-plant cooperatives, and private firms are less flexible.

The share of the four largest dairy marketing cooperatives in the total sales of cooperatives was 28 percent in 1929-31, 21 percent in 1937, and 19 percent in 1949. It declined a bit further in the early sixties. With the formation of the large regional cooperatives in the late sixties, the identity of the four largest cooperatives changed and their share of the market increased to 33 percent in 1970 and 38 percent in 1975 (O'Day, 1978).

The market power of full-service cooperatives, primarily the larger organizations, derives from--

o efficiencies in handling and routing bulk milk which allow them to perform that function at lower cost than individual handlers;

o increased manufacturing capacity for butter, powder, cheese, and other hard products which allows them to obtain over-order payments for "giving up" the potential profits from manufacturing;

o not sharing with other producers the profits from manufacturing operations including "give-up" charges.

That power is limited by the ability of limited-service cooperatives to undercut the over-order charges of full-service cooperatives. (See also Christ, p. 287.)

IN SUMMARY

Market power in packaged fluid milk and ice cream has shifted from processors to supermarket groups. In manufactured dairy products, power has increasingly moved to the packagers and distributors of branded products, who are less likely to manufacture the basic product. Large firms are increasingly leaving basic manufacturing, leaving it to cooperatives and smaller proprietary firms who often produce under contract with the larger firm.

In the markets for raw fluid milk, large full-service cooperatives have become increasingly important but limits are placed on their power by competition from limited-service cooperatives and proprietary firms who can avoid the function and costs of balancing the market.

12
Performance in
the Dairy Economy

The results or outcome of industry operations as conditioned by Government actions constitutes performance. The basic question is: How well does the industry in question meet the needs of consumers and the goals of society? This is a broad question with many aspects. It cannot be answered on a 10-point scale from "good" to "bad." But some light can be shed. Clearly it is the performance of the total system as strongly influenced by public policy and programs that is at issue. That performance has many dimensions—at least one for each function of the system, although not all can be measured and for others no measurements have been made.

Since not all buyers have the same wants and needs regarding product and price, the general level of satisfaction with the performance of a market can be expected to increase as diversity of both product and prices increase, at least up to some point. As with most other measures which attempt to provide some insight into performance, the optimal level of performance is somewhere near the middle of the range of possible outcomes. It seems likely that a level of satisfaction in relation to any possible measure of performance is U-shaped (inverted). More innovation is generally to be desired compared with less. Too much innovation will result in markedly higher costs for a very large number of products, containers, and services, and costs would rise. Up to some point, lower marketing margins reflect efficiency and indicate better performance. Beyond that point, marketing firms would go bankrupt at such a rate that consumers could not obtain the product. In short, too much of anything is not a good thing. Boulding (1980, p. 10) calls this the principle of maximum goodness... every virtue becomes a vice if you have too much of it.

Wherever possible, comparisons are provided for each performance measure to assist in putting in perspective dairy industry performance in that dimension. One cannot, of course, judge the acceptability of dairy industry performance simply by such comparisons. The fact that the price of milk has risen more or less than prices of other products is not conclusive evidence one way or the other. A part of the perspective must include interrelationships among various measures of performance. The most obvious is the relationship between price and supply—if prices are low, supplies will be smaller and perhaps inadequate to meet the desires of consumers. Conversely, if prices are kept high by public action, supplies will be large compared to the desires of the consumers and surpluses will result.

Other policies and events are affecting performance at any time-- natural causes, other agricultural policy, macroeconomic policy, and others. Such happenings probably affected dairy industry performance more than dairy policy in the seventies and likely will continue to do so in the eighties (See Babb and Boynton, 1981, p. 282).

PRICES

Milk prices, like the prices of most other goods, have been strongly affected by general inflationary pressures in the postwar period (figure 6). After the Korean War, milk prices were fairly stable through the first half of the sixties, as were the average prices of all farm products and of industrial goods in the economy. In 1966 and to some extent in 1967, milk prices rose rapidly, primarily due to price support and Federal order actions. Overall since 1947, milk prices have risen more than the prices of all farm products and less than industrial commodities. Since 1960-64, milk prices have risen more than either of the other groups.

Most of the increase in milk prices has been due to price supports which rose $9.93 between 1955 and 1981, from $2.93 to $12.86 for milk of 3.5 percent butterfat. Prices of both manufacturing grade and fluid grade milk rose less than supports and fluid grade less than manufacturing milk, due to lower Class 1 utilization.

Change in--	Change		
	1955-65	1965-75	1975-81
	- - - - dollars per cwt. - - - -		
Support price	+.15	+4.13	+5.65
Manufacturing grade price less support	+.12	+ .15	- .57

Compared to change in manufacturing grade price:

Federal order Class 1 price	+.34	- .13	- .04
Over-order payments	+.10	+ .45	+ .08
State Class 1 price	-.13	+ .10	- .49
Class 2, 3 and 4 prices	-.10	+ .12	+ .07

Average Federal order minimum Class 1 prices rose 34 cents per hundredweight more than manufacturing milk prices between 1955 and 1965, because of the addition of orders in higher-priced areas of the country such as the South. The increase was not due to higher levels of Class 1 differentials which declined 10 percent (see table 87). After 1965, Federal order Class 1 prices increased a bit less than manufacturing milk.

Relating milk prices to the prices of goods and services purchased by farmers gives a rough approximation of the relative profitability of milk production. The prices paid index is not specific to

Figure 6

Farm and Industrial Prices

Percent of 1967

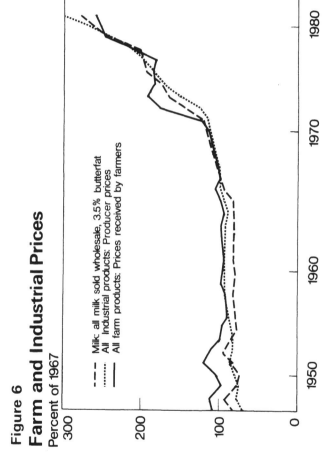

Milk: all milk sold wholesale, 3.5% butterfat
All Industrial products: Producer prices
All farm products: Prices received by farmers

dairy farmers and, of course, there is no allowance for changes in productivity. In relation to prices paid, milk prices were fairly high in the late twenties, dropped sharply during the Depression, rose sharply during World War II, and stayed generally at fairly high levels through the fifties. In the sixties and seventies, they ranged between 82 and 94 percent of old parity (calculated without regard to the ten-year-average adjustment) (figure 7).

The differential between fluid grade and manufacturing grade prices has been remarkably stable since 1950 (table 80). There was an increase in the late fifties and the most recent years show some decline. With the dollar differentials pretty much unchanged, the relative prices have changed dramatically as price levels increased. The fluid grade price was 40 percent higher than the manufacturing grade price in 1950-54 and only 10 percent higher in 1980-81. The national average Class 1 "difference" (the difference between the effective Class 1 price and the manufacturing grade price—not the same as the Class 1 differential) rose from $2.09 in 1950-54 to $2.36 in 1955-59 and stayed at about the same level in the seventies. In 1980-81, it averaged $2.64. As a percentage of the manufacturing grade price, it has declined from 79 percent in 1955-59 to 22 percent in 1980-81.

Table 80--U.S. average milk prices 1/

Period	Fluid grade less manufacturing grade	Class 1 less manufacturing grade
	Dollars per hundredweight	
1950-54	1.30	2.09
1955-59	1.48	2.36
1960-64	1.35	2.29
1965-69	1.33	2.26
1970-74	1.33	2.44
1975-79	1.26	2.32
1980-81	1.16	2.64
	Percent of manufacturing grade price	
1950-54	40.0	64.3
1955-59	49.5	78.9
1960-64	43.5	73.9
1965-69	34.3	71.1
1970-74	24.3	44.6
1975-79	14.1	26.0
1980-81	9.6	21.8

1/ Milk of 3.5 percent butterfat content.

Figure 7
Milk Prices as Percent of Parity
Percent

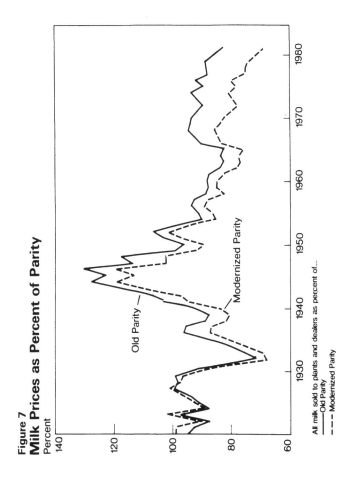

All milk sold to plants and dealers as percent of...
——— Old Parity
– – – Modernized Parity

The Class 1 difference as calculated above is made up of all or part of several components:

o Intramarket transportation.
o Intermarket transportation.
o Added costs of producing milk meeting the sanitary standards for fluid use.
o Added costs of marketing milk for fluid use (as compared to milk used solely for manufactured products).
o Price discrimation as a result of (1) Federal order or State minimum prices or (2) cooperative over-order payments.

These elements are discussed in more detail in Chapters 8 and 14.

In an analysis which aggregated all of these elements except the added cost of producing fluid grade milk, Song and Hallberg calculated the set of prices which would maximize consumer welfare and those which would maximize producer welfare in each year since 1960. Those maximizing consumer welfare had very small Class 1 differences, while those maximizing producer welfare had relatively large Class 1 differences. Given the relatively inelastic demand for fluid milk products compared to manufactured dairy products, one would expect that prices which would maximize producer returns from a given quantity of milk would involve considerable price discrimination. Their analysis showed that actual prices were closer to consumer-welfare-maximizing levels than to those maximizing producer welfare and that they moved closer to the consumer level between 1960 and 1979. In other words, if the gross Class 1 difference (which includes price discrimination and the other elements mentioned above) were all price discrimination, there was less in 1979 then in 1960.

Over-Order Payments

Federal order class prices are minimum prices to be paid by processors. At times in the past and at present, the prices actually paid by processors to producers or cooperatives—especially for Class 1 milk—have generally exceeded Federal order minimum prices. During the early years of World War II, before many Federal orders switched to formula pricing, the prices actually paid often exceeded the minimums, sometimes by substantial amounts. By 1943, the regulatory structure had been brought up to date and subsidies were paid to hold down prices. Over-order payments disappeared.

In the late forties, there were almost no over-order payments, a few during the Korean War in some Southern markets, and then the beginnings of the present developments in this field in the mid-fifties.

In a few markets, the prices established by State regulation were higher than those of the Federal order, so the effective prices were those prescribed by the State. This was true in Philadelphia from 1956 through 1965 and in Mississippi and Louisiana through most of the sixties. This provided a bonus to in-state producers supplying the market as compared to those in adjacent states.

In most markets, incentive payments were made by processors for a number of years to encourage producers to shift from can to bulk tank handling. Such payments were very common in the late fifties. They disappeared during the sixties as conversion was completed. Some individual handlers have paid premiums for certain quality characteristics, especially low bacteria count, for many, many years. Neither bulk tank premiums nor quality premiums are considered here.

Beginning in 1956, negotiated or announced over-order payments began to play a part. They started in Southern Michigan in 1956 and in Chicago in 1957. Nineteen markets--28 percent of the total--accounting for 31 percent of the Class 1 milk in the Federal order system had over-order payments in 1956, but they continued for the entire year in only two markets. One of the two was Philadelphia where the payments were required by State regulation. Most of the 1956 over-order payments were for a few months to change the seasonal price pattern. Seasonal declines had been suspended in most Federal order markets in the spring of 1956. Over-order payments were negotiated in 19 markets to maintain prices in the same seasonal relationship after July 1956 as had previously existed.

Payments above Federal order minimums have existed since then and have become more common and widespread. By 1960, the number of markets that reported over-order payments in June had grown to one-third of the total (tables 81 and 82). The percentage of markets reporting such payments held relatively stable for the next six years as the total number of markets peaked and then began to decline through mergers and consolidation of marketing areas. By 1966, slightly over one-third of the markets reported over-order payments. The proportion of markets having over-order payments increased to 65 percent in June 1970, declined to about one-third in June 1972, and increased to over 90 percent in June 1975, where it has been since then.

Although the average payments for the entire Federal order system showed some variation, there was considerably more variance between regions (table 83). During the late sixties, over-order prices were much larger in the Great Lakes region. They dropped back in line with the other regions by 1968-69, while the Southwest became the area with the largest payments. During much of the 1966-72 period, over-order payments were minimal or nonexistent in the Northeast, Florida, and the West.

In 1973-75, the over-order payments situation became much more variable and they became more widespread. Nearly all markets had some type of over-order payment at one time or another.

Feed prices rose rapidly in 1973 and 1974 as an aftermath of the Russian grain sale and an overall tight market situation for feed-grains. The Government was unwilling to raise milk prices in response to sharply escalating feed costs and over-order payments rose to unprecedented heights, although effective prices (Federal order minimum prices plus over-order payments) actually dropped. When Federal order minimum Class 1 prices rose in 1975, over-order payments declined sharply. In 1976-78, they were at approximately the same level as in early 1974.

Table 81--Frequency Distribution of Premiums Paid on Class 1 Milk in Federal Order Markets in June of each year, 1956-74 1/

Cents per Hundredweight	NUMBER OF MARKETS																		
	1956	1957	1958	1959	1960	1961	1962	1963	1964	1965	1966	1967	1968	1969	1970	1971	1972	1973	1974
1-10	1	1	0	0	0	5	2	3	2	2	5	8	2	0	2	2	4	4	0
11-20	0	3	2	3	3	5	3	4	4	5	8	6	2	8	1	1	7	10	4
21-30	0	0	8	4	8	1	4	7	4	2	8	4	9	9	6	5	4	8	4
31-40	2	1	2	1	5	3	5	3	2	1	1	4	2	2	9	7	3	1	7
41-50	0	4	1	2	9	4	2	5	5	2	1	5	23	18	10	8	2	0	5
51-over	0	3	0	2	4	6	12	8	11	11	1	3	2	3	12	15	2	2	26
Total number of Federal Order Markets Reporting Premiums	3	12	13	12	29	24	28	30	28	23	24	30	40	40	40	38	22	25	46
Total Number of Federal Order Markets	68	68	74	77	80	81	83	82	77	73	71	73	73	67	62	62	62	61	61
Percentage of Federal Order Reporting Premiums	4	18	18	16	36	30	34	37	36	32	34	41	55	60	65	61	35	41	75

1/ Premiums only; excludes service charges.

Table 82--Frequency distribution of over-order payments
on Class 1 milk in Federal order markets in June of
each year, 1974-81 1/

Cents per hundredweight	Number of markets							
	1974	1975	1976	1977	1978	1979	1980	1981
1-20	9	10	12	11	17	10	4	6
21-40	12	17	17	24	23	9	10	6
41-60	11	3	16	8	4	15	15	8
61-80	17	17	1	0	0	11	4	14
81-100	0	6	1	0	0	0	5	4
101-120	3	1	1	0	0	0	3	0
121-140	0	2	0	0	0	0	3	3
141 or more	1	1	0	0	0	0	0	3
Total reporting over-order payments	53	57	48	43	44	45	44	44
All markets	61	61	52	47	47	47	47	47
Percent with over-order payments	87	93	92	91	94	96	94	94

1/ Includes premiums and service charges.

Over-order payments are regarded by those who like to keep
things simple as purely the fruits of market power on the part of
cooperatives. And clearly there is some relationship. Without some
market power, cooperatives in many situations would have considerable
difficulty in obtaining prices above Federal order minimums. But
this is not the whole story.

In order to understand over-order payments, one must consider
not only the growth in size of cooperatives which is often equated
with market power but also the very significant changes in the functions performed by cooperatives. These were discussed in Chapter 5.
Many cooperatives are now performing functions that were previously
carried out by processors. The costs of these functions must be
recovered from processors or covered by producers. If the costs
were to be borne entirely by producer members of the cooperative,
especially in a situation where the cooperative was competing with
other cooperatives or with noncooperative producers, the cooperative
would be extremely reluctant to perform the functions, preferring to
leave them with the handler. Since, as we have seen, there are substantial economies of size in the balancing function, the processor
in many cases is not unwilling to reimburse the cooperative for
performing them if the resulting cost to the processor is less than
he would incur if he did it for himself. This often results in milk
being available to the processor at a lower cost and the processor

Table 83--Weighted average Class 1 over-order premiums,
by regions, 1966-1974 1/

Year	Great Lakes: markets 2/	North Central: markets 3/	Southwest : markets 4/	Southeast markets 5/	Average 6/
			-cents per hundredweight- - - - - - - - -			
1966	42	4	1	12	20
1967	85	20	20	29	46
1968	47	27	50	28	37
1969	31	19	50	36	29
1970	29	13	49	42	26
1971	31	27	45	36	32
1972	26	22	9	28	22
1973	41	34	51	44	40
1974 7/	71	40	74	76	61

1/ Premiums only; excludes service charges.
2/ Southern Michigan, Eastern Ohio-Western Pennsylvania, Ohio Valley, and Indiana Federal order markets.
3/ Minneapolis-St. Paul, Des Moines, Kansas City, St. Louis, and Chicago Federal order markets.
4/ Oklahoma, North Texas, Central Arkansas, San Antonio, and South Texas Federal order markets.
5/ New Orleans, Louisville-Lexington-Evansville, Nashville, and Mississippi (except in 1974 when Mississippi was not regulated).
6/ Average of these markets only. Other markets in the Northeast and West had few, if any, premiums.
7/ January-September only.

having rid himself of a number of activities which did not generate much in the way of profits.

Since 1976, pay prices to producers by the major cooperatives in each market have averaged a bit below Federal order blend prices except in one year when they were very slightly higher. If patronage refunds to producers were included, they would average a bit above Federal order blend prices. In other words, the expenses of these major cooperatives--probably all of which are full service organizations--are almost as large as the over-order payments which they are collecting. Exceptions to this general near-equality of expenses and over-order payments occur in Southeastern Florida. In the Chicago and Upper Midwest orders, major cooperative pay prices often exceed order blend prices because of the profitability of manufacturing hard products at Federal order Class 3 prices. (See Capponi, 1982.) The over-order payments in these calculations include some on

Class 2 and Class 3 milk in some markets. These are not large in the aggregate but, reflecting the give-up costs of cooperatives with major manufacturing operations, they are a significant problem to other handlers wishing to buy milk for manufactured product uses, often in competition with the cooperatives.

The geographic structure of Federal order Class 1 prices in the area east of the Rocky Mountains has been based explicitly on Eau Claire-plus-transportation since the early sixties. Before that time, the general principles had been followed but with many more variations. Distance from Eau Claire "explains" 89 percent or more of the variance in Federal order Class 1 prices in every year since 1966 (Babb et al, 1979, p. 14). Effective prices, including both the Federal order minimum and over-order payments, follow the same pattern. In such a case, the intercept value of a regression equation relating either minimum or effective Class 1 prices to distance from Eau Claire provides an average measure (not the conventional mean but a measure derived from regression analysis) of the size of over-order payments. Such a measure is provided by Babb and associates. This measure--shown in the last column of table 84--shows over-order payments of 9 cents per hundredweight in 1965 and 1966, increasing to 20-25 cents from 1967 to 1974, and ranging between 17 and 44 cents between 1975 and 1980. The reasons for the higher levels since 1967 have already been discussed: in taking over the functions of market supply and balancing, cooperatives incurred higher costs and they had sufficient market power in the upper Midwest (the base zone) to recover them. Their power to obtain such over-order charges was limited by the availability of alternative sources of milk from limited-service cooperatives, proprietary handlers, and independent producers. (See also Christ, 1980, pp. 286-7.) Some indication of the checks on collecting over-order charges is provided by comparing the charges which the major cooperative sets out to obtain and that actually received marketwide. If the cooperative had a monopoly, one would suppose that it would come close to collecting the over-order charge from everyone. In 24 Federal order markets with data available for January-August 1982, cooperatives were able to obtain at least 80 percent of the over-order charges in half of the markets. In three markets, they obtained less than half.

Babb and colleagues analyzed the data on over-order payments from 1965 through 1980 in the Federal order markets east of the Rockies to determine the factors related to over-order payments (Babb et al, 1979). The usual assumption has been that cooperative market power as measured by cooperative concentration is the major or only factor determining the size of over-order payments. Their analysis found this not to be true. Cooperative concentration--whether measured by the market share of the four largest cooperatives or the one largest cooperative--was not significantly related to the size of over-order payments. They included a number of other variables, many of which showed significant relationships to the size of over-order payments. The coefficients for measures of price relationships among Federal orders were highly significant and accounted for much of the variation in over-order payments. Significant

Table 84--Differences between the Minnesota-Wisconsin
manufacturing grade milk price and intercept values estimated
for minimum and effective Class 1 prices

Year	M-W price 1/	Difference between M-W price and intercept for		Effective difference less minimum difference
		Minimum price 2/	Effective price 3/	
	:- - - - - - - - Dollars per hundredweight - - - - - - - -			
1965	3.25	0.48	0.57	0.09
1966	3.80	.63	.72	.09
1967	4.01	.62	.84	.22
1968	4.12	.88	1.13	.25
1969	4.36	.89	1.09	.20
1970	4.63	.90	1.13	.23
1971	4.80	.86	1.12	.26
1972	4.99	.87	1.12	.25
1973	5.89	.91	1.12	.21
1974	7.26	.84	1.05	.21
1975	7.22	.92	1.09	.17
1976	8.59	.89	1.20	.31
1977	8.48	.88	1.15	.27
1978	9.29	.88	1.20	.32
1979	10.77	.88	1.32	.44
1980	11.67	.88	1.25	.37

1/ M-W price lagged 2 months to conform with procedure used to
determine Class 1 prices.
2/ Intercept values for minimum Class 1 price relationships less
M-W price.
3/ Intercept values for effective Class 1 price relationships
(including over-order payments) less M-W price.

Source: Babb et al, 1979, p. 17, and update.

differences among intercept shifters for orders and for years sug-
gested that factors not related to concentration may affect market
power and over-order payments. For example, it appeared that an
index of barriers to movement of raw milk was somewhat related to
intercept shifters for orders, but the specific barriers could not
be identified. The impacts of other economic variables were minor.
They concluded that "these findings suggest that market forces are
adjusting prices in Federal milk orders as one would expect in spatially separated markets."

Price Alignment

In a free-flow system of milk markets, economic location theory tells us that prices would be lowest in the major surplus production area (surplus in terms of needs for fluid products) and higher in other markets by the cost of transportation from the nearest market with adequate supplies available. The major surplus region in the United States has been in the upper Midwest for most of this century. Thus, in many of these analyses, we take the West North Central region which includes Minnesota as the base.

The prices of fluid grade milk reflect a number of factors. Besides the differences in Class 1 prices reflecting more or less perfectly the geographic alignment expected on the basis of location theory, they reflect differences in butterfat content, utilization in fluid products and in manufactured products, and differences in competitive conditions. Perhaps the most striking feature of these price differentials during the postwar period is their relative stability (table 85). Prices in the South Atlantic region are higher in every period except 1948-49 and 1950-54 than in any other region. Differentials in the Northeast are now about what they were at the beginning of the period. Those in the East North Central region have moved from negative to positive. They have declined in both of the South Central divisions and in the West. The decline on the Pacific Coast is particularly striking, reflecting the growth of the dairy industry in the area.

Class 1 prices which are reflected in part in the fluid grade prices are shown in table 86. These are available for a longer period, although on different bases.

Some indication of the effects of Federal order Class 1 pricing policy is given by calculating the differences between order minimum Class 1 prices and Minnesota-Wisconsin manufacturing milk prices (or comparable values for earlier years when the M-W is not available). In order to adjust for the expected differences due to spatial economics, 50 cents per hundredweight was added to the M-W price in all years and the result divided by mileage from Eau Claire, Wisconsin. The resulting figures are by no means perfect measures of order Class 1 price differentials but they have the virtue of being consistently constructed in all years and regions. They show the relative advantage to different regions and changes therein over time.

Class 1 differences so calculated were generally higher in the Midwest than in other regions (table 87). In 1960 and 1965, the Northeastern economic formulas raised them above the Midwest and in 1975 the Northeast was again higher. The South and West are below the Midwest except for 1947. The use of mileage from Eau Claire in the West is undesirable but no alternative basing point is apparent.

Federal order pricing practices were far from the only influence on actual prices. Over-order payments to cooperatives vary among regions, as previously discussed. Most State controls attempt to provide an advantage for local producers, either through higher Class 1 prices or higher Class 1 utilization or both. Most State-regulated prices were significantly higher than Federal order prices for markets east of the Rockies at comparable distances from Eau Claire in 1953-54, 1957-58, 1960-61, 1964-65, and 1975 (U.S. Agric. Mktg.

Table 85--Differences in blend prices of fluid grade milk by region

Period	New England	Middle Atlantic	East North Central	West North Central	South Atlantic	East South Central	West South Central	Mountain	Pacific
			- - - - - - - - - - cents per hundredweight - - - - - - - - - -						
1948-49	122	50	-3	0	143	105	148	66	66
1950-54	95	29	-12	0	150	95	157	76	74
1955-59	122	54	-8	0	158	87	135	80	66
1960-64	115	51	3	0	157	96	108	70	59
1965-69	98	47	10	0	140	96	110	57	15
1970-74	117	64	17	0	160	84	124	54	4
1975-79	112	53	21	0	158	76	121	52	32
1980-81	1/	1/	28	0	182	92	126	96	40

1/ Average for New England and Middle Atlantic regions combined is 92 cents.

These figures ignore differences in butterfat content between regions. They are the differences in average prices of fluid grade milk from those in the West North Central region.

Table 86--Dealers' average buying (class 1) price for standard grade milk, 3.5
percent butterfat, for use in fluid products--Differences in regional averages
from average price in West North Central Region, five-year averages, 1920-1972

Period	New England	Middle Atlantic	East North Central	South Atlantic	East South Central	West South Central	Mountain	Pacific
			- - - - - - - - - - - - dollars per hundredweight - - - - - - - - - - - -					
Prices f.o.b. shipping point or country plant, simple averages:								
1920-24:	.98	.38	.11	1.14	.19	.53	.03	.45
1925-29:	1.25	.57	.27	1.29	.24	.25	.16	.35
1930-34:	1.04	.52	.09	.88	.19	.04	.18	.21
1935-39:	.86	.57	.07	.77	.18	.10	.01	.02
1940-44:	.98	.69	.18	.89	.46	.29	-.04	.26
1945-49:	1.27	.85	.13	1.22	.66	.74	.27	.45
Prices f.o.b. city, weighted averages:								
1935-39:	.99	.74	.12	.70	.22	.07	-.08	.09
1940-44:	.94	.82	.10	.79	.39	.19	-.23	.28
1945-49:	1.17	1.06	.03	1.14	.62	.65	.17	.32
1950-54:	1.30	1.32	-.10	1.68	.78	1.31	.95	.56
1955-59:	1.81	1.75	.02	1.82	.81	1.30	1.32	.83
1960-64:	1.73	1.64	.26	1.67	1.05	1.19	1.06	1.04
1965-69:	1.34	1.03	.16	1.21	.82	1.21	.76	.28
1970-72:	1.27	.93	.09	1.06	.61	1.11	.54	-.26

Basic prices from the Fluid Milk and Cream Report, U.S. Dept. of Agriculture, last
published in June 1973.

Table 87--Federal order Class 1 price differences,
by regions, 1947-81 1/

| Year | Region | | | | United States 2/ |
	Northeast	Midwest	South	West	
	----- cents per hundredweight per 100 miles ------				
1947	10.1	10.4	12.7	--	10.4
1950	16.4	16.9	16.0	--	15.4
1955	17.5	19.1	17.4	13.5	16.9
1960	19.1	17.7	16.2	12.0	16.2
1965	17.5	16.3	15.9	11.2	15.2
1970	20.0	25.4	17.2	12.7	18.8
1975	17.0	15.8	14.0	9.8	14.2
1980	18.0	20.4	15.8	10.8	16.2
1981	20.3	26.4	18.0	12.4	19.3

--No Federal order markets in the region.

1/ Simple averages of Federal order Class 1 prices delivered to city plants less Minnesota-Wisconsin price plus 50 cents per hundredweight (3.5 percent butterfat) per hundred miles from Eau Claire, Wisconsin. Figures for some earlier years when fewer markets were regulated were calculated from percentage change in comparable markets. Small markets not included.
2/ Simple average of regions. 1947 and 1950 by percentage change in comparable regions.

Serv., 1955; Herrman and Smith, 1959; Butz, 1962; Lasley, 1965 and 1977). In a few markets with concurrent regulation by both Federal and State agencies, the State prices were significantly higher than the Federal. The State-set prices applies only to milk producers within the State (since they could not regulate interstate commerce), so handlers were tempted to go "abroad" in search of cheaper milk. In addition, blend prices were usually higher than in nearby Federal order markets that were carrying the reserve.

Retail Prices

Retail prices of dairy products have increased more slowly than the average for all foods since 1935--the increase for dairy products was less than that for the entire consumer price index or all food at home (figure 8). In other words, retail dairy product prices contributed a bit less to inflation than did other foods on the average and the average for all items. Fluid milk prices have

Figure 8

Consumer Price Indexes

% of 1967

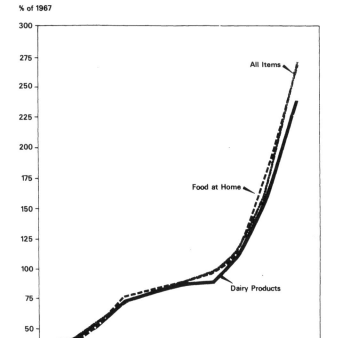

risen more slowly than those of canned milk, butter, processed cheese, and ice cream.

MARKETING MARGINS AND PRICE SPREADS

Marketing margins or price spreads measure the gross return to the marketing system to cover the costs of and profits from assembly, processing, and distribution. Technically, price spreads and marketing margins are somewhat different measures of the same general phenomenon. Farm-retail price spreads measure the difference between the retail price of a given commodity and the farm value of the amount required to yield, say, a half gallon of milk at retail at a given point in time. Appropriately measured, marketing margins would allow for the time required to move a product through the marketing system. The only measures available are of price spreads.

The farm-retail price spread for a market basket of U.S.-farm-produced foods responds to many of the inflationary pressures that affect the entire economy. Over the entire postwar period, the farm-retail spread for the market basket tracks very closely with the consumer price index for all nonfood items. Year-to-year changes may be quite different, but the trend is very similar. Price spreads for a market basket of dairy products including fluid milk, butter, cheese, canned milk, and ice cream tracks the market basket of all farm foods through the sixties. In the seventies, price spreads for dairy products increased much less than for all farm foods. The percentage increase in price spreads for dairy products was smaller than for the overall market basket in 10 of the 14 years since 1967. In 1981, price spreads for dairy products were 124 percent above the 1967 level and for all farm foods, 163 percent higher.

Margins for fluid whole milk sold in half gallon paper containers in stores increased 134 percent between 1954 and 1981 (table 88). The shifts to larger containers, to store from home delivery, and changes in price differentials meant that average margins for all fluid whole milk increased more slowly--109 percent over this 27-year period.

In 1954, the average price for milk in all containers sold in stores was a little higher than the price in half gallons, because a substantial proportion was sold in quarts at higher prices. By 1966, the average store price in all containers was below that of half gallons because of a decline in quarts and a sharp increase in the proportion in gallon containers. By 1981, these shifts had pulled the all-container price 1.7 cents per half gallon below the half gallon container price. The declining share of home-delivered milk reduced the differential between store prices and all-outlet prices from 1.8 cents per half gallon in 1954 to 0.4 cent in 1981. Thus, changes in containers and distribution methods made possible a substantial holddown on marketing margins and consumer prices of milk. These changes were also related to substantial gains in productivity in the fluid milk industry which will be discussed in a later section.

Margins for procurement and for retailing increased more than those for processing and distribution of whole milk from 1964 to 1981.

Table 88--Prices and margins for whole milk, United States, 1954-81

Year	Farm price	Transportation and other charges	Plant buying price	Retail prices				Farm-to-retail margins			
				Half gallon paper in stores	All containers			Half gallon paper in stores	All containers		
					Home delivered	Store	Total		Home delivered	Store	Total
				Cents per half gallon							
1954	21.0	1.7	22.7	42.3	47.0	42.5	44.8	21.3	26.0	21.5	23.3
1955	21.2	1.7	22.9	42.5	47.8	42.8	45.2	21.3	26.6	21.6	24.0
1956	22.0	1.7	23.7	43.8	48.8	44.1	46.2	21.8	26.8	22.1	24.2
1957	22.6	1.7	24.3	45.5	50.0	45.8	47.6	22.9	27.4	23.2	25.0
1958	22.4	1.7	24.1	45.7	51.6	46.1	48.4	23.3	29.2	23.7	26.0
1959	22.5	1.7	24.2	46.1	51.6	46.5	48.6	23.6	29.1	24.0	26.1
1960	22.6	1.7	24.3	46.5	52.7	47.0	49.2	23.9	30.1	24.4	26.6
1961	22.3	1.7	24.0	46.5	52.7	46.7	49.1	24.2	30.4	24.4	26.8
1962	22.1	1.6	23.7	46.2	52.8	46.6	49.0	24.1	30.7	24.5	26.9
1963	21.9	1.5	23.4	46.1	52.2	46.5	48.7	24.2	30.3	24.6	26.8
1964	22.0	1.5	23.5	46.4	52.4	46.5	48.6	24.4	30.4	24.5	26.6
1965	22.1	1.6	23.7	46.3	52.6	46.8	48.5	24.2	30.5	24.7	26.4
1966	24.1	1.6	25.7	49.0	54.8	48.9	50.7	24.9	30.7	24.8	26.6
1967	25.6	1.6	27.2	50.7	57.2	50.6	52.4	25.1	31.6	25.0	27.2
1968	26.9	1.6	28.5	52.6	60.1	52.5	54.3	25.7	33.2	25.6	27.4
1969	27.7	1.7	29.4	53.9	62.5	53.8	55.6	26.2	34.8	26.1	27.9
1970	28.6	1.8	30.4	56.3	64.9	56.0	57.8	27.7	36.3	27.4	29.2
1971	29.3	2.0	31.3	57.9	66.4	57.5	58.9	28.6	37.1	28.5	29.6
1972	29.6	2.1	31.7	58.8	67.2	58.2	59.4	29.2	37.6	28.6	29.8
1973	33.9	2.4	36.3	64.4	72.6	63.6	64.6	30.5	38.7	29.7	30.7
1974	40.9	2.5	43.4	77.4	85.4	76.4	77.2	36.5	44.5	35.5	36.3
1975	40.4	2.6	43.1	77.5	85.4	76.4	77.1	37.1	45.0	36.0	36.7
1976	44.4	2.7	47.0	81.8	89.5	80.6	81.1	37.4	45.1	36.2	36.7
1977	44.1	2.8	46.9	82.9	90.6	81.6	82.1	38.8	46.5	37.5	38.0
1978	47.0	2.9	49.9	87.7	95.3	86.3	86.7	40.7	48.3	39.3	39.7
1979	53.7	3.1	56.8	97.9	105.4	96.4	96.8	44.2	51.7	42.7	43.1
1980	57.9	3.3	61.2	104.9	113.0	103.3	103.7	47.0	55.1	45.4	45.8
1981	61.8	3.6	65.4	111.7	--	110.0	110.4	49.9	--	48.2	48.6

Function	Percentage change in margin		
	1964-72	1972-81	1964-81
Procurement	57	86	93
Processing and distribution	18	47	73
Retailing	-11	146	119

(Jones, 1982, and Natl. Comm. on Food Mktg., 1966.)

Local Fluid Milk Markets

Marketing margins for fluid whole milk vary widely among local markets. In an intensive study of the late sixties (Manchester, 1974a), margins were analyzed for 150 markets. To compare margins among markets where the proportions of home delivery and store sales varied widely, the margins were standardized. For this measure, marketing margins for whole milk sold on home delivery routes and through stores were weighted 30 percent home delivery and 70 percent store in all markets. The margins varied widely between markets, even after this standardization was carried out (table 89). Marketing margins were above the U.S. average in 47 percent of the markets and below in 53 percent. In 82 percent of the markets with resale price control, margins were above the U.S. average, compared with 36 percent of the markets without resale price control.

In a similar comparison using the actual proportion of home delivered and store sales in each market, the following percentages of markets were above the average: controlled markets, 68 percent; uncontrolled markets, 25 percent; all markets, 36 percent.

The effects of a substantial number of structural and institutional variables on marketing margins were analyzed for 144 U.S. fluid milk markets. Nearly 40 percent of the variation in marketing margins (standardized to remove the effects of differing proportions of home delivery and store sales) can be explained by five structural and four institutional variables plus per capita consumption (table 90). The last three columns give some idea of the effect on marketing margins of each variable at three levels. For example, the effect of market size on margins—with the effects of all other variables constant—would be to lower margins 0.02 cent per half gallon in the smallest market, 2.21 cents in the largest, and 0.44 cent in an average-sized market.

In larger markets, marketing margins tended to be lower. This is as expected, since larger markets in general have more plants serving them and there is some tendency for competition to be more intense.

The more highly concentrated markets—where the four largest firms controlled a larger percent of sales—had higher margins than the less concentrated markets. This conforms with the hypothesis that margins can be expected to be lower where more firms share more equally in the business.

Markets with larger shares of the total volume handled by integrated dairy store firms tended to have lower margins than those markets without such firms. Since dairy stores tended to have lower

Table 89--Standardized marketing margins for whole milk,
150 markets, 1967-69 1/

Percent of U.S. average	Markets with resale prices--		All markets	
	Controlled	Uncontrolled		
	Number			Percent
60-64.9	0	1	1	0.7
75-79.9	0	4	4	2.7
80-84.9	1	9	10	6.7
85-89.9	1	16	17	11.3
90-94.9	1	25	26	17.3
95-99.9	4	17	21	14.0
100-104.9	5	17	22	14.7
105-109.9	14	11	25	16.6
110-114.9	7	6	13	8.7
115-119.9	5	4	9	6.0
120-129.9	0	2	2	1.3
All markets	38	112	150	100.0

1/ 1967-69 average margins with the proportion of store and home delivery sales the same in all markets. Periods of price wars eliminated.

prices and margins than supermarkets (Jones, 1973), average margins in such markets would be expected to be lower.

The effect of isolation is measured by two variables in this analysis. The effect of distance of potential competitors from the market center is measured by the "distance" variable, while that of more or less complete isolation is measured by the "isolation dummy." In both cases, the effect of increasing isolation was to raise marketing margins.

In general, the effects of the institutional variables on marketing margins are greater than the effects of the structural variables. The most highly significant variable was price setting under State regulation. This indicates that, all other things being equal, in markets where prices were set under State regulation and had been for 20 years, marketing margins averaged 2.4 cents per half gallon higher than in other markets.

Setting of minimum prices, with market prices above the minimums, also raised prices above the levels in unregulated markets. The existence of open dating regulations in a few markets was associated with lower margins for reasons which are obscure.

The existence of restrictive sanitary regulation in a market was one of the more significant variables affecting marketing margins.

Table 90--Relationship of marketing margins to structural and institutional factors, 1967-69 1/

Independent variable	Unit	Regression coefficient	Significance (t)	Effect on marketing margins with independent variable at its--		
		Cents per half gallon		Minimum	Mean	Maximum
				----Cents per half gallon----		
Structural variables:						
Market size	Mil. lbs. per month	-.004746	1.5	-.02	-.44	-2.21
Concentration	Pct.	+.022146	1.4	+.32	+1.04	+2.21
Dairy stores	Pct.	-.032118	1.1	0	-.18	-1.01
Distance	Miles	+.004942	1.4	+.01	+.40	+.89
Isolation dummy	Dummy	+1.166660	1.4	0	--	+1.17
Institutional variables:						
Price setting	Index	+.011272	4.4	0	+.49	+2.37
Minimum price setting	Index	+.025418	1.9	0	+.07	+5.34
Dating	Index	-.007242	1.4	0	-.12	-1.52
Sanitary regulation	Index	+.117960	3.5	0	+.26	+11.80
Other variables:						
Per capita consumption	Lbs. per year	-.018457	3.3	-3.45	-5.04	-7.77

-- = Not applicable.
1/ R^2 = .3956; Standard error of estimate 2.34 cents; 144 markets

There was a tendency for marketing margins to be sharply higher in markets with highly restrictive sanitary regulations. Only a few markets were still in this category.

There was also a significant relationship between per capita consumption and marketing margins. In general, as per capita consumption rose, marketing margins declined, although the causal relationship presumably runs in the opposite direction. In other words, with lower margins there was a tendency for lower prices and consumption was higher than it would otherwise be.

INNOVATIONS

In a study of innovations in packaged fluid milk products, we have in some sense adopted the test of the marketplace. A new product or other innovation that survived in the marketplace would be deemed to be judged acceptable by society. By the same token, one that did not survive in the marketplace would be considered unaccepted and therefore unacceptable. A list of new packages, products, and services introduced in the fluid milk business since the mid-1930s was developed. Packages include all of the containers now in use except the quart glass milk bottle, which predates this period. New products include homogenized milk, lowfat milk, and half-and-half. Changes in services include quantity discounts on home delivery, store label milk, dairy stores, less-frequent home delivery, and reduced service on wholesale delivery. All of these are changes in container, product, or service which have been introduced in a number of markets and have persisted. They were, in fact, innovations when they were made.

Innovativeness was measured in terms of the average date of adoption of each of these innovations. 1/ Information was obtained from a wide variety of sources, including reports of the Federal order market administrators, research reports, The Fluid Milk and Cream Report, a survey by the Northeast Regional Dairy Marketing Committee, trade magazines, Glass Container Manufacturers Institute, a 1950 survey by the North Central Regional Dairy Marketing Technical Committee, a B.L.S. survey (1951), and a survey of processors by Lynn Fife of the University of Vermont. For each innovation in container, product, or service, the date of first introduction into the market was recorded. These figures provided the data from which an index of innovativeness was computed.

The index was computed using a set of weights developed by principal components analysis. This methodology provides a set of weights which maximizes the variance in the resulting indexes (Waugh, 1967). A wide variety of other weighting systems could have been used and undoubtedly would have yielded somewhat different results.

1/ Measuring innovativeness in this fashion ignores attempted innovations that failed the test of the marketplace. If it were possible to obtain information on these, another dimension could be added to the measures of performance. The acquisition of such data was far out of reach of the resources available for this study.

Since the objective was to derive a set of measures of innovativeness that could be related to other factors, including market structure and institutions, a set of indexes that yielded maximum variance would appear to be potentially the most useful.

The adoption dates of some of the earliest innovations, such as the paper quart, go back into the thirties. It is apparent that such early dates would have a considerable effect on the index, but it is not apparent that the fact that a market was an innovator in the thirties would continue to be significant in the sixties and seventies. Therefore, three different indexes were constructed. The first utilized the actual date of introduction, whenever it occurred. The second ignored all dates earlier than 1945, by changing them to 1945. The third ignored all dates earlier than 1950. All three indexes were used in various analyses. Only the second index, ignoring all dates earlier than 1945, is used in the analyses reported here, since the use of that index provided somewhat better measures of relationship in the analyses.

Markets ranged considerably in the rate of adoption of innovations. The market with the highest index was Milwaukee, Wis., with an average date of introduction of all innovations of 1955. The market with the lowest index was Burlington, Vt., which had an average date of introduction of 1964. These are simple averages of all of the dates of introduction of the various innovations and are not strictly comparable to the index numbers which are calculated using different weights. It is apparent that markets with resale price control were somewhat less innovative than those without (table 91).

Nearly 60 percent of the variance in innovativeness as measured by our index is "explained" by the effects of structural and institutional factors and per capita consumption (table 92). The most significant variables, in statistical terms, are isolation, price setting, and concentration.

Among the structural variables, three tend to retard innovativeness and two to encourage it. Markets with a large number of plants --which tend to be the larger markets--are somewhat slower to innovate than those with fewer plants. More highly concentrated markets are also relatively slow to innovate, as are isolated markets. Those markets where integrated supermarket chains are more important and those with a higher proportion of sales by national and regional firms are likely to innovate more rapidly than others, perhaps because larger firms are more attuned to trying new products and services (and have the resources to do so) than are smaller firms.

All of the institutional variables except labor restrictions tend to slow the rate of innovation. These include all of the regulations imposed by State and local authority including price setting, minimum price setting, restrictive licensing, restrictive sanitary regulation, and trade practice regulation. Markets with a larger number of practices restricted by labor contracts tend to be more innovative, which is counter-intuitive to say the least.

Not considered explicitly in the analysis because of the problem of quantification are the effects of other regulations including product standards. Clearly, a state or locality can inhibit innovation in products and some have undoubtedly done so. The Federal government can also inhibit product innovation, although one would not

Table 91--Index of innovativeness for markets with and without resale price control, 158 markets

Range of index of innovativeness 1/	Markets with resale price control--			All markets
	At present	Some time since 1950 but not now	Not since 1950	
	Number of markets			
+30/39	1	1	11	13
+20/29	2	1	16	19
+10/19	2	--	25	27
+01/09	2	3	21	26
-01/09	3	1	16	20
-10/19	9	3	9	21
-20/29	5	2	7	14
-30/39	9	--	4	13
-40/50	3	2	--	5
Total	36	13	109	158
	Percent			
+20/39	8.3	15.4	24.8	20.3
+01/19	11.1	23.0	42.2	33.5
-01/19	33.3	30.8	22.9	25.9
-20/39	39.0	15.4	10.1	17.1
-40/50	8.3	15.4	--	3.2
Total	100.0	100.0	100.0	100.0

-- = None.

1/ The larger the positive index the earlier the average date of introduction of the innovations in a market. All dates earlier than 1945 changed to 1945.

Table 92.--Relationship of innovativeness to structural and institutional factors, 144 markets 1/

Independent variable	Unit	Regression coefficient 2/	Significance (t)	Effect on innovativeness with independent variable at its--		
				Minimum	Mean	Maximum
		Index		Index		
Structural variables:						
Plants	Number	-0.12524	2.6	-.63	-7.32	-21.54
Concentration	Pct.	-.48649	4.3	-7.01	-22.81	-48.65
Supermarkets	Pct.	+.46383	2.2	0	+2.89	+12.48
National and regional firms	Pct.	+.17594	2.0	0	+5.95	+14.99
Isolation	Miles	-.23654	5.9	0	-17.67	-42.58
Institutional variables:						
Price setting	Index	-.12465	7.3	0	-5.47	-26.18
Minimum price setting	Index	-.30455	3.6	0	-0.85	-63.96
Licensing	Index	-.06614	2.6	0	-1.04	-13.89
Sanitary regulation	Index	-.43365	2.2	0	-.95	-43.37
Trade practice regulation	Index	-.01944	2.7	0	-2.57	-12.25
Labor restrictions	Number	+.95356	1.4	0	+2.13	+7.63
Other variables:						
Per capita consumption	Lbs. per year	+.07717	2.0	+14.43	+21.05	+32.49

1/ R^2 = .5950; Standard error of estimate = 14.15; range of innovativeness index -50 to +38.
2/ The larger the positive index, the earlier the average date of introduction of innovations in the market.

expect the results to show up in a cross-market analysis.

Markets with higher per capita consumption of fluid milk tended to have somewhat higher rates of innovation. Perhaps markets which are more ready to innovate in products and services tend to encourage somewhat higher consumption of milk products.

PRODUCTIVITY

Productivity measures for components of the dairy sector are available only for selected segments and these measures are only for labor productivity. Measures of total factor productivity, the preferred measure, are not available since 1954 (for the earlier period, see French and Walz, 1957). Even for labor productivity, the only series constructed on a basis which conforms to theory is for fluid milk processing and distribution.

The Bureau of Labor Statistics series for fluid milk processing and distribution shows large gains in output per employee hour, compared to the average for all manufacturing and for many other food processing industries (Persigehl, 1979, and U.S. B.L.S., 1979). Output per employee hour increased at an annual average rate of 4.1 percent from 1958 to 1978, with some slowing in the last five years to 3.1 percent per year. These gains reflect the rapid shift toward fewer and larger plants and changes in technology which substituted machines for human input. The overall reduction in labor also reflects, of course, the virtually complete shift away from home delivery, but the figures have been adjusted in an attempt to account for this.

Partial measures of labor use in milk production, which are not adjusted for changes in the output mix...primarily from Grade B to Grade A milk...also show substantial gains in productivity at that level (see Chapter 5).

SUPPLY-DEMAND BALANCE

Surpluses

The question of the balance between supply and demand of a given product, in this case milk, is a matter of considerable moment in any industry, but especially important for an agricultural commodity. In the absence of Government programs, price will rise or fall to clear the market for whatever supply is forthcoming. In such a situation, by definition, a mismatch between supply and demand is impossible. However, with this kind of price support program, a price floor is established by the operations of the support program and supply in excess of the quantity that will clear the market at the floor price is acquired by the Commodity Credit Corporation. Thus, CCC acquisitions provide an indicator of performance of the entire sector as it operates with public participation through Government programs. CCC acquisitions of dairy products provide an indicator of surpluses at the support price level.

These purchases by CCC have already been examined in discussing public programs (Chapter 10). When the temptation to raise support levels too high has been resisted, surpluses have been small or nonexistent. At other times, high support levels, especially when maintained for a number of years, have provided sufficient incentive to producers to increase production so that sizable surpluses were created. Lagged responses to higher producer prices, because of the biological lag, have created surpluses several years after support prices were raised.

Reserves

A reserve of milk is required for each fluid milk market, either within the market or available to it, in order to meet fluctuating demands both seasonally and in the shorter run. Otherwise, either some customers will not be able to obtain milk to meet their requirements at the going price or prices would rise sharply to ration milk. Either would be an unacceptable situation for sellers in the market, since a processor or retailer who acquired the reputation of not being a reliable supplier of milk at reasonably stable prices would be threatened with loss of sales and perhaps bankruptcy.

But carrying reserves is not costless. For farmers, reserve milk brings only the surplus-class price. Such milk can either be used in manufactured products and receive a return comparable to that paid for manufacturing grade milk or it can go down the drain and receive no return at all. There are also costs for whatever firm handles the surplus milk, whether it be a cooperative or a proprietary handler. This provides significant incentives to producers and handlers to avoid carrying reserves and to shift that role to some other group, if possible.

The voluntary nature of Federal orders--producers in any market (usually acting through their cooperative) can decide whether or not to come under a Federal order--has meant that those who anticipate a benefit from Federal regulation have generally opted for it, while those who perceive greater benefits from State or no regulation have opted to stay out of the Federal order program. Federal orders, being prohibited from preventing the entry of any producer in the supply area to participation in the pool (although there may be some costs of entry because of seasonal base plans, etc.), have carried the reserves and surpluses over and above needed reserves not only for the market as defined by the Federal order but also for nearby smaller Federal order markets and for State-regulated markets. In such cases, the secondary markets (whether Federally regulated or State-regulated) have usually maintained substantially higher Class 1 utilization than the major Federal order market. Such secondary markets have often practiced some type of exclusion or restriction on entry and on increased production by producers already serving the market.

This is a classic case of conflict between two objectives. The objective of freedom of choice conflicts with that of equal treatment of those equally situated. The growth in coverage of the Federal order system over the years has removed many of the "pockets of privilege" which formerly existed but some still remain. Outside of California, Hawaii, and Alaska, it is safe to say that every State-regulated market is relying on Federal order markets near or far to maintain all or a part of the needed reserves. While there are many fewer Federal order markets than there were in the early sixties, there are still instances where large markets carry the reserves for nearby small markets.

Interregional Shifts

The principle of comparative advantage indicates that production of a commodity such as milk can be expected to concentrate in areas where the combined costs of production and marketing are lowest, if institutional factors including public programs do not interfere. In such a situation, one would expect production of milk for use in fluid products to be located near the consuming market and production for manufactured products to be located in the areas of low production costs.

Changes in regional location of milk production were reviewed in Chapter 5. Over the past fifty years, production has declined in the Plains and Corn Belt because of competition from other agricultural products that are more profitable in those areas (table 18). The largest increases were in the Lake States and on the Pacific Coast and, to a lesser extent, in the Northeast. These are the areas with a comparative advantage in milk production. While costs are somewhat higher in the Northeast than in the other regions because of the need to import grain from the Corn Belt States, milk production increased there because of the lack of alternatives for farmers.

Differential shifts in milk production and fluid milk consumption between regions create new patterns of utilization. The proportion of milk used in manufactured products generally has increased in regions where production is rising faster than the national average and declined where the opposite is true (table 93). The biggest increases in the share of milk utilized in manufactured products since 1968 were in the Pacific and Northeastern regions.

DAIRY FARMERS' INCOMES

There are no comprehensive statistics on incomes of dairy farmers. Therefore, we have used figures from farm account records compiled by the Extension Services in New York, North Carolina, and Wisconsin and from typical farms in the Northeast and Wisconsin compiled by ERS for the earlier years. These figures have been used to estimate returns for the average commercial dairy farm in the United States, using returns per cow in these states and the average number of cows on commercial dairy farms throughout the country. The Wisconsin figures represent better-than-average commercial dairy farms in the concentrated production area in the upper Midwest. The New York figures represent the Northeast and, to some extent, other areas dependent on purchased feed concentrate supplies. North Carolina data provide an indication of returns in the South.

Dairy farm earnings are measured here as returns to operator and family labor, management, and equity. This is the total return from the farm to the operator before income taxes. Those dollars are available for family living, investment, or other purposes, along with nonfarm income. These figures are the most comparable to family income in other segments of the economy, although they are far from

Table 93--Milk used in manufactured dairy products as a percent of marketings 1/

Region	1949	1959	1964	1968	1978	1980
	- - - - - - - - - - Percent - - - - - - - - - - -					
Northeast	26.8	26.4	30.3	31.0	39.2	42.5
Lake States	74.7	75.7	77.0	74.9	78.9	83.2
Corn Belt	68.6	59.7	60.2	61.7	60.1	60.6
Northern Plains	73.0	74.1	74.0	74.8	63.3	81.0
Appalachian	51.0	42.3	46.0	42.1	40.7	39.8
Southeast	14.3	18.1	20.5	20.9	22.9	22.4
Delta States	39.4	33.8	32.9	34.0	29.8	30.0
Southern Plains	37.6	26.2	25.5	32.0	33.3	35.2
Mountain	61.5	60.4	56.3	55.5	54.6	56.8
Pacific	44.3	36.4	36.5	39.0	49.0	53.5
United States	56.3	52.2	53.6	52.8	55.4	58.8

1/ Whole milk equivalent of butterfat in manufactured dairy products as percent of whole milk equivalent of milk and cream sold by farmers. Includes Alaska and Hawaii, not shown separately, 1964-80. Regions as in figure 1.

Source: 1949-68: Miller, 1969.

perfect measures. They do not include capital gains, principally in land values, which have been significant in the postwar period.

In 1976, returns on New York dairy farms ranged from $10,000 on farms with 31 cows to $64,000 on farms with 200 cows (fig. 9). Wisconsin dairy farmer returns ranged from $7,200 on farms with 24 cows to $49,000 on farms with 142 cows. Investment on these farms ranged from $130,000 to $635,000 in New York and from $80,000 to $414,000 in Wisconsin.

In the thirties, incomes from dairy farming were low. Net incomes from dairy farming were positive only because of nonmoney income from farm-produced food and fuel and the value of housing. In 1935-36, dairy farm incomes averaged $817 per year, while money incomes for all U.S. families averaged $1,464 and total incomes (including nonmoney components) $1,784.

During World War II, incomes of all groups rose sharply with dairy farmers closing much of the gap (fig. 10). By 1945, dairy farm incomes were 86 percent of the national average money income, compared to 54 percent in 1939 (table 94). Through the mid-sixties, incomes in other parts of the economy rose faster and dairy farm incomes dropped from 96 percent of average family income in 1945-48 to 66 percent in 1960-64. In the seventies, dairy farm incomes

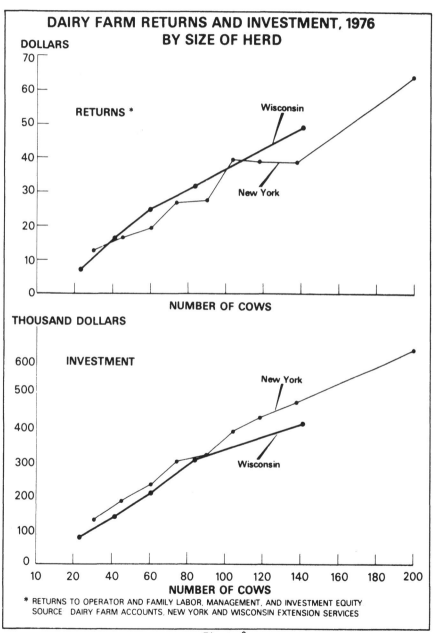

Figure 9

230

Figure 10
Incomes of Dairy Farm Families and All U.S. Families

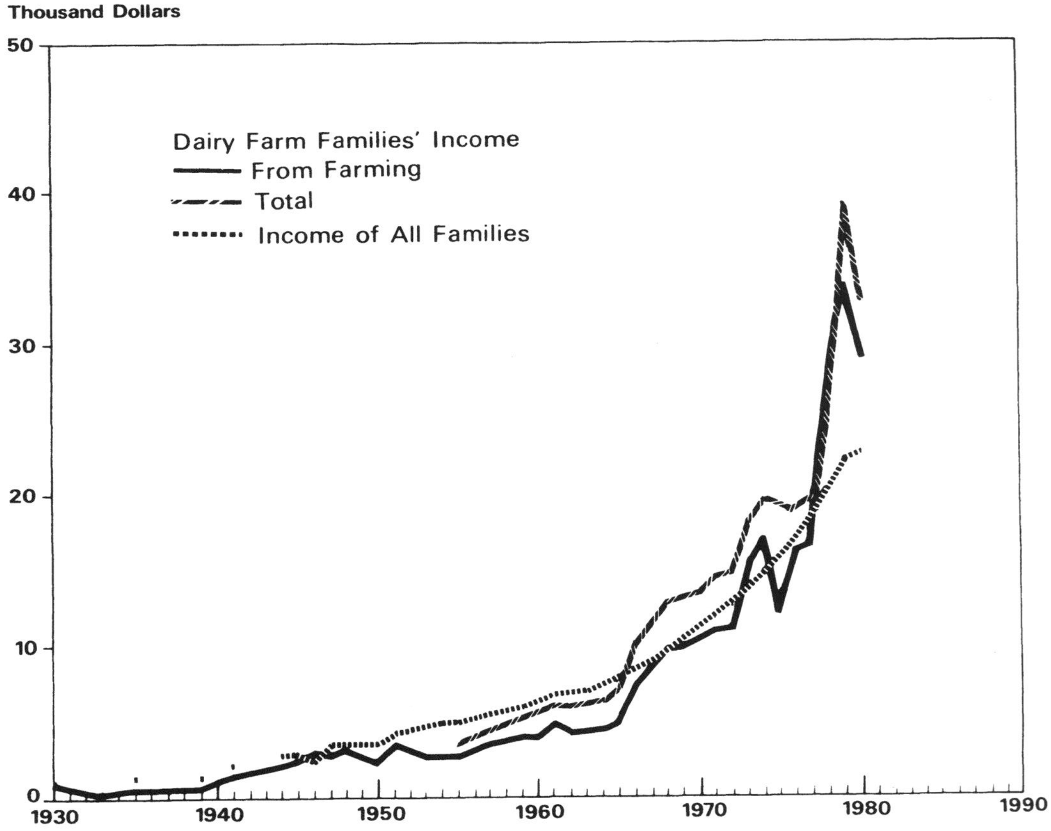

Thousand Dollars

Dairy Farm Families' Income
⎯⎯ From Farming
⟋⟋⟋ Total
▪▪▪▪▪▪ Income of All Families

50

40

30

20

10

0

1930   1940   1950   1960   1970   1980   1990

Table 94--Dairy farm incomes and comparisons

Year	Income of dairy farm families 1/ from--			Money income of all U.S. families 2/
	Farm	Off-farm	Total	
	- - - - - - - - - dollars - - - - - - - - -			
1930-34	567	--	--	--
1935-36	817	--	--	1,464
1939	720	--	--	1,329
1941	1,346	--	--	2,131
1944	2,249	--	--	3,080
1945-48	3,073	--	--	3,198
1950-54	3,123	917	--	4,309
1955-59	3,653	1,550	4,570	5,415
1960-64	4,430	2,763	5,980	6,740
1965-69	8,241	2,763	11,004	9,029
1970-74	13,246	2,853	16,099	12,729
1975-79	21,877	2,115	23,991	18,629
1980	29,200	3,588	32,788	23,974

-- Not available.
1/ Money and nonmoney income.
2/ These are the only years before 1945 for which figures are available.

exceeded average incomes.

In addition to returns from farming, many farm families receive income from off-farm employment and other sources. Comparing family incomes from all sources, the average income of dairy farm families has exceeded the average for all American families since the late sixties.

One major motivation for farm programs during the Depression and for many years afterwards was the intention to close the income gaps between farm families and others. These figures--imperfect as they are--indicate that the gap had been closed by the late sixties. The issue of equity between dairy farm families and others is no longer a policy concern, although its recurrence certainly would be.

Further consideration of the relative income levels of dairy farmers and others involves questions of economic efficiency more than of equity. The ideal of fair play in this country is more nearly "equal pay for equal work" than "equal pay for all." In such a case, the composition as well as the level of income is important. The incomes compared here include wages and salaries, return on investments, and all other sources of income. Dairy farmers have much larger investments then the average family--in 1972, about two-thirds larger on the average, although this understates the difference because of the marked concentration of wealth in the hands of the very rich. Omitting returns on investment, dairy farm incomes (including off-farm earnings) were 92 percent of the U.S. average in 1972 and probably were about equal in 1980.

FLUID MILK PROCESSORS' RETURNS

The only figures which separate fluid milk processors from other dairy processors are those of the Economic Research Service. They are based on quarterly reports to a cost-comparison service and included 70-80 plants in 1954-64 and 30 plants from 1972 to date. The firms selected are considered to be representative of moderate-size, single-plant fluid milk processors. Individual processors provide cost and sales data to a cost-comparison service to which they subscribe. That organization then provides data for selected distributors to USDA for analysis. Each processor uses a uniform cost accounting system, so the data are comparable among processors and over time (Jones and Lasley, 1980; Lasley et al, 1979; Jones, 1981).

Net returns to owners before income taxes ("net margins") were 5.0-5.6 percent of sales in 1952 and 1953, 4.0-4.1 percent in 1954-55, and 3.4 percent in 1956-61. They dropped in 1962-64 to 1.8-2.5 percent of sales. The series resumes in 1972 during price controls when net margins were 1.2 percent of sales. Since then, they have been...

1973.....1.9 percent
1974.....1.6
1975.....3.7
1976.....3.0
1977.....1.9
1978.....3.0
1979.....2.4
1980.....2.8
1981.....2.0

STABILITY

It is abundantly clear that one of the major objectives of Federal programs relating to the dairy industry is to provide a greater measure of stability than would otherwise exist. A number of dimensions of stability are involved, including:

o A measure of long-term income stability for producers. This has been approached by relating minimum prices through price supports to a measure of the prices of inputs (parity, for example).

o A measure of stability in seasonal price relationships. This has been done through Federal order classfied pricing, so that class prices and, to some extent, blend prices in the flush production season are above market-clearing levels and those in the short season are below market-clearing levels.

    o A measure of stability in spatial price relationships, so that intermarket differences in prices reflect the cost of transportation and the varying supply and demand conditions in individual markets.

    o A measure of price stability in competitive relationships-- the behavior of processors as milk buyers--so that processors equally situated pay more nearly the same class prices for milk and producers equally situated receive similar prices.

The relationship between stability and orderly marketing seems to be obvious.

It seems beyond argument that the price support and order programs have provided some measure of stability in each of these dimensions--more than would have existed without them...although one would hardly characterize milk production and prices as stable during the seventies (see Babb and Boynton, 1981, pp. 280-3). Some have argued most vigorously that they have achieved that stability at too-high a price level. Others have argued equally vigorously that the price and income levels achieved have been inadequate at times to provide a decent level of living for dairy farmers and to ensure the long-term capacity to produce milk in quantities sufficient to meet consumer demand. The question of the appropriate level is deferred for later consideration. Here the question of the effects of price stability is addressed.

It is clear that in the seventies the price support program has been addressed primarily to the question of the overall level of milk prices and dairy farm incomes, while the emphasis of the Federal order program has been on the other dimensions of stability. Both Federal orders and price supports played an income-support role up through the sixties.

The effects of the price support program have been to modify substantially the risk of bankruptcy for dairy farmers over the past forty years. By providing a floor under all milk prices at the low point of the price cycle, supports have kept prices to farmers above a level of out-of-pocket losses most of the time. There have been times, as in 1973-74, when feed prices were very high and out-of-pocket losses were fairly general among dairy farmers. But most of the time such extreme conditions have not existed and cash flow has been positive.

An unwanted side effect of this stability is that there is some tendency to keep less-efficient producers in business by aborting the "wringing out" of marginal operators. Evidence of the importance or unimportance of this effect is lacking. Somewhat offsetting is the increase in efficiency because farmers are enabled to make greater investments in more efficient buildings, equipment, and cows owing to greater certainty as to returns.

Farmers in other lines of animal agriculture have never had and apparently never wanted comparable price support programs. All livestock products are subject to a price and production cycle, because of the biological lag in production. These cycles are of different duration and amplitude because of both differences in the biological characteristics of production and the existence of the dairy price support program, which modifies the cycle in the case of milk. The price patterns for cattle feeding and hog raising are markedly different than that for milk (figure 11). Since 1964, steer prices rose in eleven years and fell in six, while hog prices—with a shorter cycle—increased in eight years and decreased in nine (table 95). Milk prices, on the other hand, did not decline in any of those years, although the increase varied from 0.6 percent from 1976 to 1977 to 17.6 percent from 1972 to 1973. The magnitude of price variations was significantly larger for steers and hogs. (See also Mayer, 1979.)

Table 95—Year-to-year percentage change in prices of livestock products, 1965-1981

Years when prices—	Choice steers at Omaha	Barrows and gilts at 7 markets	All milk sold to plants and dealers
	NUMBER OF YEARS		
Increased	11	8	17
Decreased	6	9	0
	AVERAGE PERCENTAGE CHANGE		
Increased	12.8	29.4	7.3
Decreased	4.4	9.8	---
All years	9.1	12.8	7.3
	LARGEST PERCENTAGE CHANGE		
Increased	29.6	51.0	17.6
Decreased	12.3	17.5	---

It is also clear that the Federal orders have provided a substantial measure of stability in the other three dimensions described above. The problem of inter-seasonal variations in prices is treated in a generally satisfactory manner through classified pricing and pooling. The instability engendered by the seasonal shift from short supply to surplus no longer causes short- or long-run instability.

Spatial price problems are of a different magnitude and character in recent years compared to the early years of the Federal order program, because of technological developments in milk transportation and preservation. Distant markets are now much more closely related than they were. This, of course, increases the importance of stability in the spatial dimension, as well as the need for more frequent

Figure 11
Farm Prices of Cattle, Hogs and Milk
Dollars per CWT

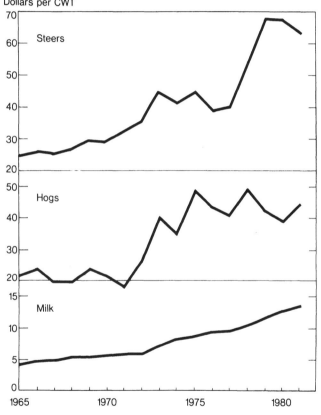

adjustments. It seems clear that Federal orders have made a significant contribution in providing spatial price stability and that performance in this dimension can be improved.

The installation of a Federal order which requires all processors similarly situated to pay the same class prices for milk made a
major change in the dimension of competitive stability. For many
years thereafter, the advantage which one processor could gain over
another, if both were regulated by an order, did not include the
ability to pay less to farmers for milk for the same use. In the
last fifteen years, with the transfer of major functions from processors to cooperatives and the institution of over-order payments to
cover the cost of these services (and perhaps more), the opportunity
to gain such an advantage has returned.

Stability has value to consumers as well as producers. Dairy
product prices at retail have been more stable then those of other
animal products, in part because of the operation of the programs.

SOCIAL COSTS

The growth of the regulatory reform movement in the seventies
brought a widespread use of the tools of welfare economics to measure
social costs. Milk was a natural subject for such analyses for several reasons, including (1) the fact that milk is perhaps the most
regulated food commodity, (2) the fact that classified pricing has
the potential for price discrimination in the economic sense, (3) the
availability of rigorous theoretical work on the possibilities of
price discrimination in individual markets, and (4) the ready availability of data on prices and quantities. There was a flowering of
such studies. A sample of the discussion includes Masson and Eisenstat, 1978 and 1980; Ippolito and Masson, 1978; Gordon and Hanke,
1978; Alagia, 1979; Buxton and Hammond, 1974; Dahlgran, 1980; Fones
et al, 1977; Roberts, 1975; Masson et al, 1978; Buxton, 1977 and
1979; Dobson and Buxton, 1977. The numbers for social costs are
almost invariably large in such analyses but small relative to the
totals. Costs in the millions translate to less than 1 percent of
consumer expenditures for milk.

One of the key concepts in such analysis is that of economic
surplus. The additional amounts that consumers would have been
willing to spend to acquire (smaller) quantities of milk, in this
case, beyond that which they in fact paid is defined as consumer's
surplus. It is measured by the area under the demand curve and
above the price line. In analogous fashion, producer's surplus is
defined as the difference between the price actually received and
the price at which producers would have been willing to produce
(smaller) quantities. If the effective prices are changed by regulation, the change in each of the measures of economic surplus is
completely defined by the relative elasticity of the demand and supply schedules. Given the state of the arts in measurement of supply
and demand elasticities, it is almost the universal case that the
curves are defined in linear fashion either in arithmetic or logarithmic terms. Because of the great convenience provided by a constant measure of elasticity when logarithmic transformations are
used, measurements in these terms are much more common.

But one of the great truths of economic analysis is that we can only provide approximate measures of the magnitude of elasticity (or any other economic construct) which is within the range of experience—i.e., within the range of the data available. Thus, for all the reasons familiar to those who estimate and use empirical measures of elasticity, it is seldom the case that it is possible to estimate elasticities over the entire range needed in order to compute either the consumer's surplus or the producer's surplus.

Cochrane has entered a spirited challenge to the idea of consumer's surplus (Cochrane, 1980). "The concept of consumer surplus is, for me, an illusion created by an alternative purchasing procedure which is not operational." As he said, the shape of the demand curve outside the range of experience is unknowable. Thus, a mechanical extension of the demand curve using constant elasticity is meaningless. This shortcoming is somewhat mitigated by the fact that it is the change in consumer's surplus that is of interest, eliminating the need to consider the farthest reaches of the demand curve.

The concept of producer's surplus is even more slippery. At least where the demand for food is concerned, no capital investment by the consumer is necessary to buy more or less. But for major change in the output of milk, substantial investment or disinvestment is required. There is no reason to suppose that the shape of a supply function describing the investment process is the reverse of a supply function describing the disinvestment process. Yet this is assumed.

A fundamental premise in calculations of social costs and benefits of changes in regulation is that any decrease in the measured net benefits is an indication of an undesirable change. But every such analysis must start from some existing or assumed set of institutions and distribution of wealth and power and measure change from there. Thus, one set of rights is assumed to be preferable to another but the assumption is never defended or subjected to analysis. (See Shaffer, 1979, p. 722).

This body of theory has been used to demonstrate that under perfect competition price instability leads to greater social benefits than does price stability. But this theory assumes away the problem —perfect knowledge and perfect mobility are assumed. Thus, producers and consumers are assumed to be able to forecast prices without error and to adjust to the new levels at zero cost. If these conditions do not hold, as they cannot in the real world, the results are the opposite. (See Mann, 1977, p. 123.)

A related diffculty is that such analyses assume that equilibium exists, a conclusion which follows from the assumed existence of perfect knowledge and instantaneous perfect adjustment. But such an equilibrium is never achieved.

For these and many other reasons, computations and interpretations of social cost are fraught with difficulty and subject to severe limitations. None will be found in this book.

SYNTHETICS AND SUBSTITUTES

Milk has been disassembled into its various components almost since the beginning of recorded history. It was a matter of simple observaton that, if milk was allowed to stand, the cream rose to the top. Early man learned to extract the butterfat by mechanical means and make butter or butteroil. Disassembly of the nonfat solids portion of milk was somewhat slower to develop. A higher degree of specific knowledge, especially of milk chemistry, was required. The ability to divide milk into various components creates a multitude of products which can be used as ingredients in both food and nonfood manufacturing. In almost every use, there are competing products from nonmilk sources.

Until World War II, butterfat was the principal component in terms of value, with the skim portion being purely a byproduct of relatively low value. Early competition for butter came with the development of butter substitutes that were first based heavily on oleo oil from beef fat and later vegetable oils. Colored margarine carried a substantial tax from 1902 until the early fifties but nevertheless made substantial inroads into the market for table spreads. With the repeal of the tax, the butter market shrank over the years until it is now a specialized market in higher-priced restaurants that want to maintain a quality image and in households with similar budgets and tastes. (In 1972-74, households with incomes over $25,000 spent three times as much on butter as those with incomes below $3,000--and spent more on butter than on margarine. U.S. Bu. of Labor Stat., 1977, pp. 26-7.) Butter use as an ingredient in manufactured food products is limited to those where the manufacturer desires to impart either a butter flavor or a butter image. Butter has not been used in quantity in commercial baking for many years--as long ago as 1929, butter was only 9 percent of shortenings used by bakers.

The nonfat solids portion of milk had little value in the thirties once it was separated from the butterfat. A modest amount was dried, principally for feed use. Another modest portion was manufactured into casein, primarily for use in paint and paper manufacture. Most skim milk from butter factories went back to the farm for animal feed or down the drain. Nearly all whey from cheese plants went into the sewer.

Construction of nonfat dry milk plants by the Government during World War II in order to utilize nonfat solids for human food and the support price for nonfat dry milk created a demand for the product by giving it economic value. The domestic casein industry disappeared, as it could not compete with nonfat dry milk production in the absence of a support price. The relative value of the fat and nonfat portions of milk have changed markedly over the years since World War II. In 1953, butterfat accounted for 84 percent of the value of manufacturing grade milk and, in 1981, 49 percent.

During the same period, numerous advances in food technology created new products with new uses in food manufacturing. Developments in casein and caseinates changed the uses of this fraction of milk from almost entirely nonfood to mostly food. Somewhat comparable changes in the technology of soy proteins created new products that compete with milk components in many food uses. The markets for nonfat dry milk, casein, caseinates, and other milk derivatives have changed markedly as a result.

Changes in the use for food ingredients of dry milk, casein, and whey products during the past twenty years are dramatic (table 96). Processed meat products, once a significant outlet for nonfat dry milk, use much less. Nonfat dry milk used in processed meats was down to 45 million pounds in 1969, largely as a result of the extension of Federal inspection and standards to plants which had formerly been under State inspection and standards. A number of States had allowed addition of nonfat dry milk in sausage and similar products at higher levels than permitted under Federal standards. When they came under Federal inspection in the sixties, these plants had to conform to Federal standards or obtain a special label showing how much nonfat dry milk was added and the use of nonfat dry milk in those plants dropped sharply. Since, then, the decline in nonfat dry milk usage resumed after an upsurge when calcium-reduced nonfat dry milk was introduced. A small portion was taken up by casein, whey, yeast proteins, and single-cell proteins.

Twenty years ago, the bakery market was far-and-away the most important ingredient use for nonfat dry milk. Much of that market has been lost to whey. Part of this change is due to technology rather than price. Bakers found that a "baker's mixture" composed of dry whey, sodium caseinate, and mineral salts worked better and cost less than nonfat dry milk, particularly in the continuous-mix process of bread baking which was becoming the dominant technology in the sixties.

In prepared dried mixes for cakes, rolls, and related products and in confectionery, the use of milk ingredients increased, although there has been significant substitution of whey products for nonfat dry milk.

Use of nonfat dry milk in dairy products has increased about 15 percent in twenty years, although it was higher ten years ago then it is now. At the same time, the use of whey has increased sharply, and it now appears to be larger than nonfat dry milk usage. The principal uses of dry milk and whey in the manufacture of dairy products are for fortification of lowfat and skim milk products, frozen desserts, processed cheese foods and spreads, and cottage cheese. Usage has increased for all of the products, with significant substitution of whey for nonfat dry milk in frozen desserts and processed cheese foods and spreads.

The most important food usage of casein is in substitute dairy products, including imitation cheese, whiteners, whipped toppings, and similar products. Imitation cheese, which was unknown twenty years ago, is by far the most important use.

Bulk condensed milk has not replaced dry products during this period. There are no annual data for uses similar to that for dry products, but total production of bulk condensed milk (whole, skim, and buttermilk; sweetened and unsweetened) declined by 19 percent between 1959-61 and 1978-80.

Substitutes for dairy products have taken major portions of a number of submarkets over the years. Margarine now accounts for most of the table spread market. Per capita consumption of cream and half and half is a fraction of what it was twenty-five years ago, with the major portion of the market now accounted for by coffee whiteners and whipped toppings whose only dairy ingredient is imported casein. In

Table 96--Sales of dry milk, casein and whey for ingredient
use in foods

Use	1959-61		
	Dry milks 1/	Casein 2/	Whey 3/
	- - - - - - - - - - - - - Million pounds		
Meat processing	66	--	*
Bakery	336	1	*
Prepared dry mixes	64	--	--
Blends	--	--	--
Confectionery	55	--	--
Baby food	4	1	--
Other foods	31	2	--
Dairy products:	214	1	*
Fluid fortification 4/	17	--	--
Frozen desserts 5/	58	1	*
Cheese 6/	33	--	--
Cottage cheese 7/	*	--	--
Dairy substitutes:	--	--	--
Imitation cheese	--	--	--
Coffee whiteners	--	--	--
Other	--	--	--
Total	769	5	30-50

*Some, but amount unkown.
--Not available. Usually none or amall amount.

1/ Nonfat dry milk, dry whole milk, and dry buttermilk. Source: American Dry Milk Institute.

2/ Sources: 1959-61: Rough approximations based on Miller, 1971. 1978-80: 1980 usage from U.S. Econ. and Stat. Serv., 1981b.

3/ Includes lactose. Sources: 1959-61: Rough estimate of total based on U.S. Econ. Res. Serv., 1962. Uses from Groves and Graf, 1965. 1978-80: Estimated from Whey Products Institute.

Table 96 (cont.)

Use	1978-80		
	Dry milks 1/	Casein 2/	Whey 3/
	Million pounds - - - - - - - - - - - - -		
Meat processing	16	1	3
Bakery	87	8	165
Prepared dry mixes	73	--	35
Blends	--	--	84
Confectionery	72	1	27
Baby food	2	1	77
Other foods	25	12	101
Dairy products:	246	--	313
Fluid fortification 4/	63	--	--
Frozen desserts 5/	91	--	97
Cheese 6/	77	--	29
Cottage cheese 7/	15	--	*
Dairy substitutes:	--	56	--
Imitation cheese	--	42	--
Coffee whiteners	--	12	--
Other	--	2	--
Total	522	79	805

4/ Estimated assuming that 90 percent of the solids used in fortification is nonfat dry milk and 10 percent is condensed skim.

5/ Calculated using New York State data in New York Crop Rpt. Serv.

6/ Calculated.

7/ 1978-80: Incomplete reports in American Dry Milk Institute and Whey Products Institute.

recent years, imitation cheese has grown to a significant but unknown level, primarily for use on frozen and restaurant pizzas. Tilsworth and Graf estimate that imitation cheese had taken a third of the Mozzarella market by 1980-81 (Tilsworth and Graf, 1981, p. 1). Filled milk (using vegetable fat in place of milkfat) had a surge in the sixties and could be expected to return if Federal and State regulations were changed. Ingredients of vegetable origin, both vegetable fats and soy products, are less expensive than milk products and would remain so even in the absence of regulation.

Generally sympathetic attitudes of both National and State legislatures toward dairy farmers have resulted in numerous laws and regulations inhibiting substitute dairy products over the years. The tax on colored margarine has been referred to. The Filled Cheese Act of 1896 placed similar taxes on filled cheese which used fats other than butterfat. The Filled Milk Act of 1923, originally passed to prohibit the movement in interstate commerce of filled evaporated milk, was effective for 50 years in preventing the substitution of vegetable fat for butterfat in canned milk, ice cream, and fluid milk products. Its teeth have now been drawn by court decision. Standards of identity for a number of food products were effective for many years in restraining use of substitutes for dairy ingredients. Such restrictions have become increasingly less effective in restraining the use of substitute products and can be expected to fade further. Federal order regulation has inhibited introduction of concentrated milk, filled and imitation milk, and reconstituted milk. The tendency often is to consider that these products would have been resounding commercial successes in a less-regulated market--and some might have been, but hardly all of them.

IN SUMMARY

The assessment of the performance the dairy economy with its particular mixture of private and public enterprise involves many tradeoffs--is somewhat more in a given desirable dimension worth what is given up in another dimension? If all could be reduced to a measurable common unit--the utility of economic theory--, then one could make the correct decision as to appropriate public and private policy. Lacking such a measure of the collective good, we muddle through in dairy policy as in all other policy questions by trading off stability for income or whatever. It seems likely that the tradeoffs are less appreciated than they might be and that the attention of many is concentrated on one objective, ignoring all others, so that tradeoffs do not exist for them.

There is probably a fairly general acceptance among the public of the proposition that farmers should be as well-remunerated as other segments of society for comparable work. This income parity idea has often been among the objectives of agricultural legislation. Defining what would constitute such a parity of income is an arduous, probably imposssible task. But being definitive about the direction of change is feasible: Dairy farmers' incomes have become more like those of nonfarm families since 1965, having been quite distinctly lower in earlier years. Since changes in price support levels translate fairly directly into higher gross and net incomes for dairy farmers, the choice between a price support program which provides stability and one which also significantly enhances incomes involves a judgement as to what constitutes an adequate income.

13
The Dairy Industry Today

In Chapters 5-7, the many changes in the dairy industry since the thirties were discussed in considerable detail. At first glance, it would indeed appear that nothing is the same. Dairy farming is now principally the business of 175,000 commercial dairy farms, each producing several times as much milk per person as in the thirties. But, except in Florida, Hawaii, and the Pacific Coast states, farms are still one- to three-man operations, essentially family farms.

The dominance of fluid milk processing and distribution by processors that characteriszed the thirties has been changed radically. Supermarket groups now set the pace of prices and competition in the marketing of fluid milk products, with processors much less powerful as a group than they were.

Cooperatives are larger and more powerful than they were during the Depression. In fluid milk, they have grown from single-market or area cooperatives (where several markets were closely interrelated) to regional and multi-regional in scope. The functions of balancing and managing the raw milk supply for fluid markets is now largely handled by these larger cooperatives, where it was mainly performed by each proprietary processor in the thirties.

Changes in the scope of operations of fluid milk cooperatives reflect to a considerable extent changes in the scope of markets. For raw milk, formerly local independent markets are now a part of an interrelated system of markets covering the entire country. Raw milk regularly moves several thousand miles. Similarly, but on a smaller scale, packaged milk markets have grown larger. Formerly independent markets 40 miles apart are now merged into a single market and packaged milk regularly moves 200 miles or more, with some movements of 400 miles. Thus, the competitive relationships of the sellers of packaged milk are much different.

Supermarket groups have become very different buyers, even where they have not integrated into fluid milk processing as a number of the larger groups have done. Where they formerly handled fluid milk essentially on a commission basis, they are now contract buyers of private label milk. Since these contracts are easily moved from one processor to another, processor profits are often minimal.

Processing of manufactured products is gradually moving away from the sellers of branded goods. More and more, they are contracting with others--usually cooperatives--to perform the basic processing, while they concentrate on merchandising and distribution. This shift is noticeable in hard cheese, cottage cheese, and butter.

Does all of this mean that the conditions which led to the public dairy programs in the thirties are so changed that there is no longer a reason for those programs, even though the present programs are considerably different than those conducted under the same names in an earlier period?

Many of the basic characteristics of the dairy industry which were discussed in Chapter 2 were changed little or not at all by these developments. The dairy industry still has very different characteristics from those of other agricultural and nonagricultural industries.

Production still requires specialized inputs that are of limited if any use in other activities. They are now many times more expensive than they were in the thirties. The rate of output is still controllable chiefly through variations in the number of cows being milked, which means that it can be reduced by slaughtering cows but increased only by breeding and raising more cows. Entry and exit are still expensive. Inspection of the output for quality and safety is still required.

Milk is still produced every day and must move to market at least every other day. While that is only half as often as it moved during the thirties, milk is a flow commodity and marketing and pricing must deal with that fact.

In the short run (day-to-day), supply is not tuned to demand... the cows produce and milk must go to market, even if the demand is not there on that particular day. By the same token, demand for milk for bottling is almost zero on a Sunday and small on Saturday and Wednesday, since most plants close down, but the milk is still produced and can be discarded only at a high cost. There are substantial economies of size in the management of milk supplies to deal with these day-to-day variations. A single manager can carry out the task much more efficiently than several firms. This provides a considerable incentive to centralize the management of milk movement and supplies in a market, a task which has largely been taken over by cooperatives for this reason.

This change in functions, which is largely the creature of the past twenty years, has, of course, contributed greatly to the growth of the large regional cooperatives. Since they could perform the function more efficiently than the processors to whom they sold, there was considerable incentive for both cooperatives and processors to move in that direction. The result has been that individual cooperatives often supply a high percentage of the milk in a given market... in small markets, 100 percent; in larger markets, often two-thirds or three-quarters of the milk.

In an industrial market, if one manufacturer had three-quarters of all sales, he would be well on the way to being called a "monopolist" by the antitrust agencies and at least some courts. But "control" of 75 percent of the supply of a raw agricultural product such as milk hardly confers the same market power as a similar percentage of the supply of a differentiated consumer product. The market share of the cooperative seller of agricultural products has a quite different base than is found in markets for industrial products. Manufacturers of steel or automobiles have virtually complete, legal control over output and the productive resources used to create that output. Manufacturers who consider it to their economic advantage have both the power and the right to restrict production. In the imperfectly competitive industrial markets of today and yesterday, barriers to entry on the part of potential competitors make power to restrict production a meaningful index of market power, though far from perfect.

Most agricultural marketing cooperatives lack such power. Production decisions for a given crop or product are in the hands of farmers, some--perhaps most--of whom are members of the cooperative. Many cooperatives have marketing agreements with their members which require that the cooperative market all of a member's production but which provide no means of control by the cooperative over the volume of that production.

The firm buying from that cooperative, on the other hand, is under no such obligation to accept supplies. The buying firm may specify the quantity or accept it only at reduced prices. Thus, the market power of a marketing cooperative with a market share of, say, 75 percent cannot be equated to the market power of a buyer, say a processor, who takes 75 percent of the product.

The "free-rider" problem is basically unchanged from the thirties. A limited-service cooperative or proprietary milk supplier can avoid the costs of managing and balancing supplies for the entire market, letting large, full-service cooperatives "carry the surplus" and cover the costs. With over-order charges in most Federal order markets to cover the costs of managing and balancing supplies plus the proceeds, if any, of market power, a limited-service supplier can sell at minimum order prices and effectively undercut the larger full-service supplier. This effectively provides a limit on the power of the large cooperative to exploit its apparent position of "control" of the market. Without minimum order prices, the possibilities would be even greater.

Other externalities--situations of the "free-rider" type where costs and benefits do not accrue to the same groups in society--still exist, some of them in a more heightened form than fifty years ago. All possible substitutes for Class 1 milk products have this potential. Federal order market pools that carry the reserves for other markets have this characteristic.

Thus, there remain a number of possible bases for public intervention in the marketing and pricing of milk. The alternatives are explored in the next chapter.

14
Policy Alternatives for the Eighties

The basic policy choice to be made during the next decade is how much and what kind of role public institutions should play in the dairy economy. One possibility to be considered is that, given the changes in the dairy industry since the thirties, Federal dairy programs no longer serve the public purpose and, therefore, should be abolished. In the preceding chapter, the highlights of the changes of the past 50 years have been discussed and some fundamental characteristics that remain unchanged have been laid out. The summary statement might be: everything has changed except the fundamentals.

Another possibility is that future changes in technology might cause drastic changes in milk production or marketing, making all government intervention, or at least market orders, irrelevant or counter-productive. None of the possible changes in the technology of production seem likely to cause such a change. It seems quite possible--perhaps even likely--that developments in genetics or in other aspects of dairy science will markedly increase efficiency in milk production. While this would surely shift the supply curve, it is unlikely to change the biology that causes the lag in bringing about major increases in production--i.e., the cycle is not likely to be eliminated. Nor would such changes eliminate the need for large capital investments in milk production.

In marketing, a technology that changed fluid milk from a perishable product requiring refrigeration into a dry grocery product would eliminate all of the special features of fluid milk marketing including its inherent instability. Such a technology exists in ultra-high-temperature (UHT) sterilization. But the costs are sufficiently high that UHT offers promise only for a small, highly specialized market--not to replace fluid milk (see Miller, 1982b). If the costs come down sharply to somewhere near the levels of fluid milk, the story will be different.

Reconstituted milk is sometimes promoted as performing a similar function, but it could do so only in a limited way even if it were to be permitted by changes in Federal and state regulations. The added costs of concentrating and reconstituting a fluid milk product are high enough that it would not be competitive within several hundred miles of the base zone for Class 1 prices. At greater distances from the base zone reconstituted milk would be competitive with fresh fluid milk with change in pricing practices. It's effects will be discussed later.

A caution is in order in considering impacts of potential technological change: more often than we tend to remember, vast change has been predicted. Most of the time, it hasn't happened or, if change did occur, progress was more nearly glacial than apocalyptic.

THE EIGHTIES WITHOUT FEDERAL REGULATION

The principal dairy-specific programs that could be eliminated are the price support and Federal order programs. Eliminating both price supports and market orders would remove authority to limit dairy product imports. Continuing either one but not the other would maintain the authority for import quotas. An additional dimension is provided by consideration of modification of Federal policy towards agricultural cooperatives. This, of course, extends beyond milk marketing cooperatives.

The principal effect of the price support program is to provide a floor under milk prices at the low point (in terms of prices) of the cow cycle. It can also be used to raise the income levels of dairy farmers and it has been, although the existing program is not well designed for that purpose since surpluses build up in the absence of supply controls and government costs reach high levels. In view of these problems, only the stop-loss type of price support program is considered here, with consideration of other programs deferred to a later section.

A purchase program of the present type, with support prices at minimal levels so they take effect only when farm prices of milk approach the low point of the price cycle, clearly changes the expectations of dairy farmers--compared to those that would exist without such a program. Farmers make plans and investments with the understanding that the floor under prices exists and will continue to be there. A major element of risk has been removed by the existence of such a price guarantee.

Abolition of the guarantee provided by a stop-loss price support program would change the risks facing dairy farmers. Milk prices would behave more like those of other livestock products (see table 51). On biological grounds, one would expect them to behave more like steer prices than hog prices, but without the upward shift in demand over time as consumer incomes rise.

Without price supports, the magnitude of the milk price cycle clearly would increase, as the lower limit was removed. Given the inelastic nature of both supply and demand functions, very large changes in price would be required to clear the market during periods of cyclically large supply (see Christ, 1980, p. 280). Testing with an econometric model of the dairy industry--even one that could perfectly explain all that had happened in the past, which none of them can--to determine the effects of removing price supports for milk will provide some insights, but not all the required information. (See Hallberg, 1982.) The supply function in any econometric model is estimated on the basis of historical experience. Unless one goes back to pre-1940 data, statistics on the reactions of dairy farmers to a situation without price supports are unavailable. The changes in technology and economics since 1940 are so great that it seems highly unlikely that a model based on the twenties and thirties will provide much insight into likely changes in the supply responses of dairy farmers in the eighties if price supports were removed. This is not to fault the analyses that have been made but to point out the limits to the information they can provide.

Ever since the first statistical analyses of supply functions, it has been recognized that supply response probably is different when prices are declining than it is in periods of rising prices. But the multiplicity of factors affecting output in any given year is so great that it has seldom been possible to segregate periods of rising prices from those of declining prices for analytical purposes. Thus, analysts usually implicitly assume that the supply function is reversible--i.e., that the elasticities are the same regardless of the direction of price change--even though they may suspect some divergence from that situation in the real world.

This problem hardly seems important in analyzing dairy supply functions when prices (at least annual average prices in nominal terms) have not decreased in 15 years. But, with the removal of price supports, it would clearly become much more important than now, and perhaps crucial...even though we have no data on which to base an analysis.

There can be little question that drastically changing the magnitude and character of the risk facing dairy farmers by eliminating the guarantee of a floor under prices would significantly affect their evaluations of the economic situation facing them and their reactions to it. The current price support program reduces one element of risk--that involving extremely low levels of milk prices. It does not remove other risks, such as high feed prices or loss of market for the individual producer, and it does not guarantee profits on milk production, as the out-of-pocket losses of dairymen at times such as in 1973-75 demonstrated. Increasing the frequency of such periods and their severity would unquestionably increase the incidence of bankruptcy and its near-relatives among dairy farmers. If bankrupt farmers are merely replaced by others who are able to buy into the business at lower levels of investment, the consequences for the economy are of one kind. If, on the other hand, no one can be persuaded to enter the dairy business to replace a significant portion of the bankrupt, then the consequences are indeed different. Either milk production declines sharply and permanently or existing producers with greater staying power due to larger equities or more tolerant bankers expand to replace the lost production.

Dairy farming is the least concentrated of all animal products enterprises (table 97). It is the last bastion of the family farm in animal agriculture except for hog farming, where disease problems prevented the development of large specialized enterprises until recently. The stability provided by the price support program is a major contributor to this situation. Its removal would create pressures toward large-scale organization in dairy farming, along the lines of California and Florida production, rather than Wisconsin and New York.

Under the specialized conditions now typical of California, Arizona, Texas, Florida, and Hawaii, Matulich finds significant economies

Table 97--Concentration of farm production on the largest and smallest 10 percent of all farms reporting that commodity as their major enterprise and selling at least $2,500 of all farm products in 1974

Commodity	:	Number of farms	:	Production controlled by--	
	:		:	Smallest 10 percent of farms	Largest 10 percent of farms
	:		:	(Percent)	
Most Concentrated	:				
Eggs	:	193,533		0.2	96.2
Fed cattle	:	210,725		0.3	83.5
Irish potatoes	:	33,142		---	73.6
Vegetables	:	58,580		0.2	68.4
Oranges	:	12,803		0.1	62.7
Peaches	:	10,728		0.1	59.2
Turkeys	:	4,407		0.1	55.2
Apples	:	18,953		0.3	52.7
Concentrated	:				
Cotton	:	80,724		0.5	51.6
Wheat	:	502,621		0.9	49.3
Feeder pigs	:	93,234		1.5	48.3
Beef cows	:	724,569		1.0	48.2
Least Concentrated	:				
Tobacco	:	151,017		1.3	42.6
Corn	:	857,830		1.1	41.1
Hogs and pigs	:	393,280		1.2	39.3
Soybeans	:	512,853		0.9	38.8
Milk cows	:	242,189		0.8	38.2

Source: U.S. Department of Commerce, 1974 Census of Agriculture, Volume I, Part 51, Tables 20 through 25, December 1977.

of size up to 750-cow herds (Matulich, 1978). This is drylot dairy-ing, with all feed and forage purchased. These are industrialized milk production operations, comparable in many ways to large-scale beef feeding operations. Labor is unionized.

Leaving aside questions of the economics of such drylot opera-tions in the rest of the United States, suppose that 750 cows became the average size of all dairy herds in the United States, with aver-age production of 20,000 pounds of milk per cow or 15 million pounds per farm. In such a case, 8,000 farms would produce all of the milk in the country. This is hardly the stuff of which oligopoly is made. In short, present technology in production is not about to change dairy production into something closely resembling most of manufacturing. It would not even resemble the broiler industry with its 150 contractors...who have had no success in getting the production and price cycle under control in that industry.

In other words, while such an extreme would destroy the family dairy farm with its perceived social virtues, it would not create an industry with power to control its own output and prices, as most manufacturing industries can.

The limitations to the growth of very large factory-type dairy farms in New York State are probably typical of the major Northeastern and Midwestern dairy states. In a study of nearly all of the very large farms (those with more than 150 cows) in New York State, the only major limitation on growth was not availability of capital nor technology, but the preference of owner-operators for the kind of operation they now had. This was basically the "family" farm, where the operator(s) had day-to-day participation in the business of dairying. They had either chosen that line of enterprise because they like hands-on dairying or had come to prefer it after engaging in it. None of them were interested in exchanging this kind of activity for the business of managing several hundred employees. (See McGuire and Stanton, 1979.)

In cattle feeding and producing broilers, turkeys, and eggs, the substantial risks have led either to large-scale units, as in cattle feeding, or contractual integration, as in poultry products. These are ways of dealing with the risk problem. If similar risk problems were created in milk prices, ways would have to be developed to deal with them. One would expect that the kind of risk-spreading that is only possible with large corporations might be a possibility in milk production.

Removing Federal Orders

In an interdependent system based on price supports and Federal orders, a principal function of orders is to provide a degree of stability in competitive relations. The objective is to provide more nearly equal treatment to those similarly situated. Removal of Federal orders would return fluid milk markets to a more nearly free-market situation with some minor constraints provided by antitrust activities.

The economic incentives of different types of processors and cooperatives would be substantially changed by removal of Federal order minimum prices to producers. A processor or limited-service cooperative that could avoid the costs of servicing and balancing the bulk milk market would be able to substantially undercut the price of a full-service cooperative or processor. This would likely lead to some rather nasty competitive battles with the "big boys" pitted against the "little guys." The antitrust agencies would be active. Given the characteristics of milk production, processing, and distribution (see Chapter 13), the ultimate outcome might be either monopoly cooperatives (although it is beyond belief that the antitrust agencies would tolerate them) or continued battling between numerous competitors with the attendant costs of a fragmented system. Given the inherent instability discussed in Chapter 2 and elsewhere in the book and the interests of the antitrust agencies, it seems much more likely that the market power and membership of large full-service cooperative would decline in many markets.

Under the present system, many producers switch from one handler to another--13 percent of Grade A producers surveyed had switched in five years (Boynton and Babb, 1982). The key question is: under the changed circumstances could large cooperatives retain their members or even add more? The opportunities for limited-service cooperatives and other arrangements would be much greater if the requirement of paying at least Federal order minimum prices were removed. They would have complete freedom to undercut the prices of full-service cooperatives, leaving them to carry the full costs of balancing the markets. Since there is no advantage to the individual producer in belonging to an organization which performs this essential but costly function, it seems highly likely that defections would mount as other opportunities opened up.

A contrasting view is given by Hallberg: Bulk milk can now move much greater distances than it could during the twenties and thirties and dairy cooperatives are now solidly entrenched and, for the most part, sophisticated business organizations and fairly effective bargaining or marketing agents. "Hence, assuming there is good information flow in the industry and that dairy cooperatives continue to retain their members (i.e., to be effective marketing agents), it is questionable that marketing orders are now needed to provide milk producers with the bargaining power needed to present instability (on the scale that existed in the 1920s and 1930s, at least) that might arise because of the buying practices of milk handlers." (Hallberg, 1980, pp. 3-4.) The key question is: would cooperatives retain their members? I regard it as much less likely that they would be able to do so. (See also Dobson and Buxton, 1977.)

Removing Federal orders would also remove the guarantees of honesty and accuracy provided by auditing and check-testing. Recent experience when a new Federal order was introduced indicates the potential: checking butterfat tests showed that 41 percent were inaccurate, all too low, compared to 4 percent in a nearby market where check-testing had gone on for years. The opportunities for inaccurate reporting of utilization would greatly increase without auditing.

Another casualty of eliminating orders would be an end to marketwide pooling, except in a few small markets where all producers were members of one cooperative. This would result in considerably different returns to members of different cooperatives and nonmember groups, which would hardly be perceived as equity.

These consequences are sufficiently serious to lead serious students to propose Federal orders without pricing provisions as a possible alternative.

ALTERNATIVE POLICIES--PRICE SUPPORTS

Examination of policy alternatives other than complete rejection of the Federal role in milk marketing regulation involves a variety of issues. These issues can be divided into that of cyclical stability or instability and that of competitive stability. The price support program deals with the first issue and the Federal order program with the second. Income enhancement for dairy farmers is approached through both programs--and through cooperative policy--but, in recent years, it has been done more through price supports than through Federal orders.

The basic policy choice relating to price support programs is between (1) providing--as the present program was designed to do--a measure of price stability to dairy producers and consumers by setting floors at minimal levels and (2) supporting milk prices substantially above market-clearing levels in order to provide higher incomes to dairy farmers. If the choice is to provide floors at minimal levels, a purchase program of the present type can work efficiently. If the choice is income enhancement, other programs will have to be used to restrain production, encourage consumption, or both.

A purchase program of the present type maintains prices of dairy products in the United States well above the levels at which they are available for export from major producing countries. Since all significant dairy countries operate public programs, many involving large subsidies to exports of dairy products, the resulting "world prices" should not be confused with free-market prices that might exist without such programs. In such circumstances, import controls are necessary to prevent flooding the U.S. market with foreign dairy products.

The key element in making a purchase program work is that support levels be held at minimal levels. If they are raised to keep prices above market-clearing levels over a substantial portion of the cycle, production outruns demand at the support level and substantial surpluses begin to accumulate at considerable cost to the Government. No one can say with certainty what the level of surpluses and Government costs may be which, in a particular set of circumstances, would trigger public and legislative reaction to dismantle the price support progam, but the probability that such a level exists is very high. The experiences of 1981 and 1982 are instructive. In such circumstances, prudent public policy would select another type of program if the objective of substantial income support is chosen.

Alternative programs include direct payments and supply control. A direct payment (or deficiency payment) program has many of the features of the target pricing approach to price supports utilized for major crops in the late seventies. Payments would be made directly to individual producers when prices fell below the designated support level. This kind of program separates income support from the price system. Payments could be made on all milk or only on manufacturing milk or they could be made only on quota milk. (See also Vertrees and Emerson, 1979, and Schnittker et al, 1979.)

At the same support level, a payment program has different results from a purchase program:

o Consumer prices are lower.
o There are no Government stocks to dispose of.
o Costs are shifted partly from low-income to higher-income

consumers, because prices are lower and taxes are higher. The progressive income tax structure does the rest.

o There is no effect on efficiency or on the ease of making resource adjustments, as long as supports are at modest levels.
o With high supports, supply control is needed to control Government costs.
o Government costs are higher.
o If imports cannot be controlled, a payment program is the only alternative because it lets dairy products sell at world prices.

If the basic objective is substantial income enhancement, some form of supply control is necessary to limit Government costs. Most forms of supply control use sales quotas, allocated to each producer as a percentage of historical production, with a penalty against milk sold above quota. The penalty would have to effectively reduce value of over-quota milk below the marginal cost to produce it...otherwise, farmers would have an incentive to produce milk in excess of quota and effectively defeat the purpose for creating a quota system.

Quotas would immediately be worth money, giving windfall gains to the farmers holding the original quotas. The higher milk prices would be capitalized into the quotas. Production costs would be increased for new producers and existing farmers who bought additional quota. Both the value of the quota and its effectiveness in controlling supply would be affected by how easy it was for new or existing dairy farmers to acquire quota.

Compared to a purchase program at the same level of support, a supply control program has these characteristics:

o Leaves consumer prices unchanged.
o Reduces Government costs.
o Requires detailed regulation of individual producers.
o Restricts ability to adjust resources.
o Results in capitalization of quotas.
o Has not been favored by producers in the past because of the loss of freedom.
o Without import regulation, supply control will only work in combination with direct payments.

These are more or less "pure" programs. It is possible to combine features of several programs, thus providing more tools to deal with the problems being faced by dairy price policy and avoiding some of the problems that arise with a single program. A direct payment program on manufacturing milk has the advantage over a purchase program of allowing the market prices of manufactured products to fully react to market forces, even when supplies are large. When prices are at support levels with the present program, manufactured products are less competitive with substitutes--butter with margarine and nonfat dry milk with whey and caseinates, for example. But the need to support returns to all producers at approximately the same level severely restricts the opportunity to tilt the prices for individual manufactured dairy products.

Combining a payment program with a manufacturing grade milk order would make it possible to let market prices reach market-clearing levels. At the same time, it would provide support to returns to dairy farmers regardless of the manufactured product in which their milk was used. Under such a program:

o A marketing order for manufacturing grade milk would set class prices for milk used in various manufactured products. Different classes and prices would be established for milk used in cheese, butter-powder, evaporated milk, ice cream, and other products.

o A payment would be made into the market pool by the Government equal to the difference between the class price and the target level.

o Similar provisions would be incorporated in all other milk marketing orders.

o A support purchase program could be continued with prices lower by the amount of the payment, if desired.

Such a program would have a number of effects and some problems:

o At a time of large surpluses, prices of manufactured products would be lower and consumption larger. Cheese would be more competitive with meat, butter with margarine, and powder with whey and caseinates.

o The problem of distress milk would become more crucial. Distress milk is that which is surplus to a handler's needs when milk is plentiful and other handlers are not interested. Special order provisions would be required to deal with it.

o The competitive market for manufacturing grade milk and the Minnesota-Wisconsin price series would disappear. Different methods of determining class price movers would be required, probably product prices for manufactured products and target prices for fluid products. This would make the manufacturing margin (make allowance) more crucial than at present. Regular reports of manufacturing costs would have to be required of processors, with auditing by Federal order market administrators.

While the differences in price elasticities of demand for milk used in different manufactured products are not large, such a program would remove constraints on pricing, so that sales and consumption of each product could be maximized. This would reduce Government costs somewhat, compared to a program where the price for milk is the same regardless of the manufactured product made from it. It would be an extension of classified pricing and subject to the same criticism of price discrimination.

Modifications in the Purchase Program

Interest in alternatives to the purchase program was minimal or nonexistent for many years, except occasionally among analysts, although there was a resurgence of interest in a quota program in 1982. Discussion has centered on possible modifications of the purchase program. Most of those considered revolve around changes in the products purchased and their relative prices and changes in the price standard.

Changing Relative Prices of Purchased Products. The comparative value of butterfat and solids-not-fat is effectively determined much of the time by the purchase prices for butter and powder under the price support program. Until 1970, legislation required supporting both milk for manufacturing use and butterfat in farm-separated cream between 75 and 90 percent of parity. This left the Department limited discretion in determining relative values of butterfat and solids-not-fat.

Relative prices of butter and nonfat dry milk have changed markedly, primarily as a result of price support actions. In 1960, the price per pound of butter was 4.3 times that of nonfat dry milk. The ratio stayed about 4-to-1 through 1965 and declined to near-equality in 1974. In 1981, it was 1.6-to-1.

While there are no legal limits on the relative values of butterfat and nonfat solids set under the price support program, competition from other products places considerable constraints upon the prices that are set. Butterfat is in direct competition with vegetable fats in many uses and nonfat dry milk competes with whey, soy products, and caseinates for many purposes. Since whey prices are not supported and cheese support prices are set so as to return to farmers nearly the same amount as butter and nonfat dry milk combined, whey can and does sell at much lower prices than nonfat dry milk. This will continue to present problems until either whey is brought under the price support program or nonfat dry milk is removed from it. Either of these actions would present very difficult problems.

The main impact of changing relative purchase prices of butter and powder would be on the mix of dairy products purchased under the support program. Government costs would not be greatly affected. However, products could be encouraged which were easier for CCC to give away, if there were any.

Because butter and nonfat dry milk are produced in fixed proportions from whole milk, the value resulting from their combined prices must be compared to the value of milk used in other manufactured products. The value of milk used for butter and nonfat dry milk production can be changed relative to the value of milk used in cheese production by adjusting purchase prices. Usually, the two have been set approximately equal. However, a tilt in purchase prices making milk more valuable for cheese production than for butter and nonfat dry milk production was in effect in 1973-74 and in 1976-77. Tilting the relative value of milk used for a particular dairy product would, over time, result in more purchases of that product under the current price support program. The processing sector of the dairy industry would attempt to divert surplus milk into the use with the higher value, creating adjustment problems for those plants unable to convert and for producers selling milk to them.

If tilts were to become a regular feature of price supports,

Federal order pricing policies for milk used in butter-powder and cheese would have to be reevaluated. With a tilt toward cheese, there would be a tendency for cheese plants to pay higher prices than butter-powder plants--to an even greater extent than they now do. Under present Federal order procedures, regulated plants would be paying more for milk used in butter-powder than unregulated plants and less for milk used in cheese. This created considerable strains in 1973-74 and 1976-77. If it were to become a regular practice, Federal orders would probably need to establish separate classes for butter-powder and for cheese.

Changing the Products Purchased. The Secretary of Agriculture can decide which dairy products will be purchased under the price support program. The longer-run implications of purchasing only butter and nonfat dry milk are very similar to those of tilting purchase prices in favor of butter and nonfat dry milk. The industry would adjust when milk is in surplus so that surplus milk would be diverted into the product that the Government was willing to buy. Milk would be diverted into the manufacture of that product until the value of milk was approximately equal in all manufacturing uses. The larger the number of dairy products purchased the more easily the adjustments could be made in allocating surplus milk between manufacturing plants. On the other hand, butter and nonfat dry milk have advantages since they can be used as ingredients in a wide variety of products.

It is tempting to conceive of a purchase program with only cheese supported. That would let butter and nonfat dry milk find their own price levels in the market, making them somewhat more competitive with vegetable fats, soy products, or caseinates. Cheese would become the residual use for milk. Such a program might be manageable if support levels were actually kept at stop-loss levels, but competition from filled and imitation cheese would be heightened. The benefits of such a change would be small and the costs of adjustment would be large.

Price Standard. The price standard for the support program has been stated in terms of parity since the Act was passed in 1949, when price supports for practically all other commodities used a similar parity standard. Beginning with the 1973 Act, there has been a shift for many commodities from parity to cost of production. Milk is now one of the few commodities using the parity standard. Within legal limits stated in terms of parity, the standard is adequate supply.

One of the primary movers in the parity calculation is the index of prices paid by all farmers for production inputs and family living. The weights in this index are average purchases by all farmers, regardless of the products produced. This means that the index does not move one-to-one with the prices of items purchased by dairy farmers. For example, feed has a weight of about 20 percent in the prices paid index but amounts to about 50 percent (including pasture and roughage) of dairy farm costs. Also, the prices paid index includes items not used by dairy farmers--feeder livestock and baby chicks, for example.

Milk production can be very profitable or unprofitable at any particular percentage of parity, depending on many items the most important of which is feed (table 98). The higher the parity index is relative to feed costs (last column) the more profitable dairy farming is at a given percentage of parity. These figures show why a high proportion of dairy farmers lost money in 1974, even though milk prices were over 80 percent of parity and why milk production was profitable and increased sharply in 1978-81.

The other major mover of the parity index for milk is the relative prices of milk and other farm products in the preceding ten years. This feature was introduced by the Agricultural Act of 1949 in an effort to "modernize" parity. Its effect is to raise the 1910-14 base price for all milk sold to plants and dealers, currently from $1.58 to $1.96 per hundredweight. The April 1982 parity price for all milk wholesale was thus raised from $16.85 to $20.90 per hundredweight. The all-milk price would have been at 79.5 percent of parity, if calculated by the old method, rather than 65 percent of parity by the "modernized" method.

A further element in the computation of parity prices for milk is that changes in butterfat content are ignored, even though milk has been priced in terms of butterfat content for many years. The butterfat content of milk sold wholesale (to plants and dealers) is now around 3.65 percent, varying a bit from year to year, but mostly unchanged through the seventies. This, of course, reflects the fact that most market milk now comes from Holsteins. In the twenties and thirties, the butterfat content was typically 3.92 percent and it was at least that high or higher--perhaps up to 4.0 percent--in the 1910-14 base period for parity price calculations.

The effects of the 10-year-average adjustment and butterfat pricing can be seen in calculations of parity treating these matters differently:

	Parity prices for April 1982	
	At actual butterfat test	At 3.5 percent butterfat
Adjusted	$20.90	$19.05
Not adjusted	16.85	15.36

This calculation assumes butterfat content of 3.92 percent in 1910-14.

An alternative standard is cost of production. It is affected by the prices paid by farmers and by many other factors, including the impacts of bad weather on pasture and roughage production, yields of grain on those dairy farms producing their own grain, feeding rates, and technological change. All of the factors which affect cost of production have an impact on dairy farm income, but so many factors are involved in one measure that it is not possible to sort out the temporary from the longer-lasting. Cost of production is difficult to compute and depends on many assumptions made by the analysts. For example, cost of production is figured for the dairy enterprise only, rather than for the entire dairy farm including grain production. Thus, when grain prices are high and its production is quite profitable, this will appear as a high cost for dairy

Table 98--Parity index and cost of dairy feed

Year	Parity index	Dairy concentrate ration cost	Parity index / Dairy ration
		- - - - 1960-64=100 - - - -	
1960	97.8	98.3	99.5
1961	98.4	98.3	100.1
1962	100.1	99.3	100.8
1963	101.7	102.3	99.4
1964	102.0	102.0	100.0
1965	105.0	102.0	102.9
1966	109.2	106.0	103.0
1967	111.1	108.7	102.2
1968	113.8	104.3	109.1
1969	119.3	106.0	112.5
1970	124.5	110.4	112.8
1971	130.4	115.7	112.7
1972	138.5	118.4	117.0
1973	160.0	164.2	97.4
1974	181.9	209.6	86.8
1975	199.8	210.3	95.0
1976	212.8	212.0	100.4
1977	224.6	208.6	107.7
1978	243.5	204.6	119.0
1979	277.1	224.8	123.3
1980	311.3	249.7	124.7
1981	330.2	270.9	121.9

farmers even though many of them are producing their own grain and profiting from it. Average cost of production is the measure used, but it masks a wide divergence among dairy farmers.

On the other hand, an index of prices paid by dairy farmers for inputs that they purchase is relatively free of the problems that plague computation of cost of production. It is simple to figure and thus would be available currently.

An adequate supply standard would result from removing the limits on the Secretary's authority to set price support levels. Price supports would then be set at the level necessary to bring forth enough milk to meet consumer demand without building large CCC stocks. Such complete flexibility is popular with analysts and some other groups but it is unacceptable to most producer organizations and their friends.

If Federal dairy policy is to provide stability rather than

significant income enhancement for dairy farmers, regardless of the standard used, some flexibility must be provided to the Secretary to determine the level of price support in the light of changing supply and demand conditions. None of these alternative standards (except adequate supply) reflects all the factors influencing both supply and demand, so they could not be used to set the support price directly. Permanent legislation provides a range from 75 to 90 percent of parity. If parity is retained as the price standard and stability is the objective, increasing productivity and low feed prices make it necessary to reduce the minimum percentage of parity to 65 percent or less.

One possibility is to relate the discretionary price support range to CCC purchases as was done in the Agricultural and Food Act of 1981. This would provide a narrower range for the Secretary's discretion but would relate that range to supply and demand conditions. For example, if CCC purchases were more than 3 percent of supply, the discretionary range might be 60 to 70 percent of parity. If CCC purchases were below 3 percent, the range could be 70 to 85 percent of parity. Similar flexibility could be provided with a dairy parity index. (See Manchester, 1978.)

The problems involved in the use of parity prices as currently calculated under the amended Agricultural Adjustment Act of 1938 (amended in 1948, 1949, 1954 and 1956), as previously discussed, have led various analysts to suggest a dairy parity index as a possible substitute (Manchester, 1978, p. 41; Schnittker et al, 1979, p. 32). The proposal is to utilize an index of the prices paid by dairy farmers for inputs that they purchase in lieu of the index of prices paid by all farmers for inputs and family living. This would have the index track the actual cost-prices faced by dairy farmers much more closely and would avoid the problem indicated by table 98.

One such index is shown in table 99. The dairy-specific index is higher than the all-farm index when feed prices are relatively high and lower in other years.

Utilizing the information thus developed, a possible new parity calculation could be constructed as follows:

o The base year is 1977. This is the government-wide base year for index construction. In addition, it was a year of relative supply-demand balance in the dairy industry with profitable production but without large CCC purchases or stocks.

o The parity price for 1977 is that calculated by the methods described earlier--it omits the 10-year-average-price component which merely confirms the price trends established by earlier price support levels. Also it is calculated at 3.5 percent butterfat.

o The parity price for other years is calculated by applying the dairy prices paid index to the 1977 parity price of $9.92 per hundredweight for all milk sold at wholesale (to plants and dealers).

o An appropriate adjustment to reach a manufacturing milk parity price would be to apply a three-year average differential for manufacturing grade milk prices from the all milk price in table 100--not the ten-year average now used.

Table 99—Dairy and all-farm prices paid indexes

Year	All-farm prices paid index	Dairy prices paid index	Dairy index / All-farm index
	- - - 1977=100 - - -		
1965	47	47	100
1966	49	49	100
1967	49	50	102
1968	51	49	96
1969	53	51	96
1970	55	54	101
1971	58	56	97
1972	62	59	95
1973	71	79	111
1974	81	94	116
1975	89	94	106
1976	95	99	104
1977	100	100	100
1978	108	104	96
1979	123	121	98
1980	138	136	99
1981	150	146	97

Calculating parity for milk in this fashion would be entirely appropriate and would provide an updated measure of prices that would yield adequate returns to dairy farmers. Since 1965, actual prices would have ranged from 88 to 106 percent of parity. Figuring parity on this basis, the minimum level could have been 85 percent of parity without causing any large surpluses.

An additional feature that is not included here would be a productivity adjustment. With appropriate treatment, milk per cow could be used as the basis for such an adjustment. The usual standard is that it takes three addtional pounds of grain in the form of concentrates to produce ten more pounds of milk. This standard could be applied to provide a productivity adjustment which would prevent the parity measure from getting out of hand in the long run.

These alternative standards are considered here for use in a purchase program such as the present one where support levels are selected to provide stability but not significant income enhancement. The choice of standards for an income enhancement program involves many other considerations.

Table 100--Possible new parity for milk of 3.5 percent butterfat

Year	:	New parity price	:	Price of all milk sold wholesale	
	:		:		
	:	Dollars per hundredweight			Percent of parity
	:				
1965	:	4.66		4.09	87.8
1966	:	4.86		4.66	95.9
1967	:	4.96		4.87	98.2
1968	:	4.86		5.11	105.1
1969	:	5.06		5.36	105.9
	:				
1970	:	5.36		5.58	104.1
1971	:	5.56		5.74	103.2
1972	:	5.85		5.93	101.4
1973	:	7.84		7.01	89.4
1974	:	9.32		8.20	88.0
	:				
1975	:	9.32		8.59	92.2
1976	:	9.82		9.49	96.6
1977	:	9.92		9.55	96.3
1978	:	10.32		9.79	94.9
1979	:	12.00		11.77	98.1
	:				
1980	:	13.49		12.76	94.6
1981	:	14.48		13.54	93.5
	:				

Reducing Production. From time to time, even in the best-run price support program, production can be expected to get out of line with consumption and large surpluses result. If the attempt is made to use a [purchase program] for income enhancement, such a condition will occur more often. Changes of the kind discussed in the preceding section would minimize such occasions. When supply gets out of line with demand, there is always a search for ways to get them back into line.

Perhaps the simplest, though not therefore the best, method is to [lower the price support level until balance is restored.] But, if the surplus is large, the relative elasticities of supply and demand are such that large price decreases would be required to obtain the adjustment in a year or two. The alternative is to [price the marginal output of each producer low enough to greatly encourage adjustment by the required amount,] putting the marginal return below marginal (out-of-pocket) cost. This would require a base for each producer to identify the portion of his production to which the lower excess price would apply.

A straightforward method of calculating a base would be to specify it as a fixed percentage of his deliveries in a recent period before discussion of the introduction of a base plan commenced (in order to offset increases in output in anticipation of the introduction of such bases). Administration of the plan would be greatly simplified by assigning bases to units rather than to individual producers whenever possible and letting the organizations deal with their members. These would usually be cooperatives but a proprietary plant could deal with its nonmember producers.

The chief difficulty with such a base plan is to prevent the bases from acquiring value and distorting future adjustments by producers. The objective is get a rapid adjustment and then wipe out the bases before they are capitalized. To this end, bases would be nontransferable and last only one year. If the supply adjustment were not completed within a year, the support price for all milk would be lowered to the level needed to bring supply into line with demand.

CHOICES FOR MILK MARKETING ORDERS AND COOPERATIVE POLICY

With the flexibility afforded by the Agricultural Marketing Agreement Act of 1937, the Federal order system has evolved in the past 45 years from a system treating individual markets separately to a unified system in which all markets are parts of a near-national system. But the development of such a system is never completed... further evolution is both possible and desirable. A number of major policy issues are involved in determining the direction for further evolution. Each of these issues involves tradeoffs. The time would seem to be ripe for a reassessment of Federal order objectives and functions in the light of these policy issues.

One issue revolves around the nature of the regulatory process-- is the best public policy effectuated by what is fundamentally an adversary process or by one where the element of cooperation between Government and governed is stronger?

The Federal milk marketing order program developed along lines of joint activity by the Government and milk marketing cooperatives (see Chapter 9). This kind of institution is sometimes regarded as more sophisticated and superior. It is a form of the "social contract," in which various groups in the society come together for the solution of their problems.

The Federal Milk Order Study Committee under the chairmanship of Edwin G. Nourse, former Chairman of the Council of Economic Advisers, characterized the milk market order system as "a truly unique marketing institution, neither quite free nor fully controlled but heavily 'conditioned' by both private and public mechanisms and policies" (Nourse et al, 1962, p. 6). The Committee's premise was that, in an age of monopolistic competition and administered prices, the institutions appropriate to a bygone and largely mythical age of atomistic competition are not useful. The public interest--which subsumes and compromises the interests of farmers, processors, and consumers--is represented by the Secretary of Agriculture who has a responsibility for "determining the price level and coordinating the national price structure such as no other Cabinet officer possesses" (Nourse et al, 1962, p. 87).

A diametrically opposed point of view had long been held by the antitrust community and, in the late sixties and seventies, by the

new "public interest" groups. This point of view holds that all regulation should be carried out at arm's length under the strictest of
adversary conditions. This point of view tends to rule out not only
the type of cooperation inherent in the market order system but also
the use of economic incentives to achieve the objectives of environmental or safety regulation.

Federal orders both underwrite and circumscribe the powers of
cooperatives, as E. G. Nourse pointed out twenty years ago: "Since
it undertakes to make milk marketing orderly and equitable among all
factors in a given market and between related markets, the Secretary
must decide (a) what structures and practices that have grown up
between producer and handler organizations in any given market shall
be accepted for inclusion and enforcement in a Federal market order,
(b) whether the terms of trade are properly related to the basic
price established in the order, and (c) which of them shall be modified or abandoned--whether through attrition or summary action."
(Nourse et al, 1962, pp. 8-9).

Since that time, the Federal order system has completed its
development into a near-national system of closely related markets
and the question of what to do in an individual market is much less
important than formerly, because--even with larger, regional markets
--the effects of any decision on the entire system must usually be
considered dominant. But the basic issue remains--by his actions,
the Secretary can strongly influence the amount of market power that
a cooperative has. Decisions on many topics affect the market power
of some or all cooperatives. Often, the effects on full-service
cooperatives with a large share of the market are quiet different
from those on smaller, limited-service cooperatives. Nearly all of
these questions discussed in the following sections have implications
for cooperative market power.

Voluntary or Comprehensive System

The Federal order system developed as a voluntary enterprise.
Producers in any area retain the right to say whether or not they
wish to come under a Federal order. In the early days, since orders
were applied to individual markets that were typically isolated from
each other in an economic sense, the resulting rather spotty coverage
remained workable. But the nation east of the Rockies is now one
system of interrelated markets, most of which are Federally regulated
but not all. Pockets of privilege exist between Federal orders where
producers enjoy the benefits of the existence of the order without
paying a part of the cost. Producers in most State-regulated markets
and in the unregulated areas between some Federal orders are in a
position to let the Federal order system carry the reserves and the
burden of the surplus, if any. Over the years, the number of areas
which were able to stay out of the Federal order system has declined
sharply, but there are still some.

The hearing process has provided a means of modifying orders to
keep them attuned to rapid changes in marketing. On the other hand,
the voluntary nature of the program has meant that it is difficult
for the Government to make changes until cooperatives feel they are
needed. As markets become more interrelated, it is necessary to look

at the entire system of orders when making changes. This complicates the problem of obtaining consensus among producers and requires greater leadership on the part of Government in securing producers' awareness of needed changes in the order system. It also means that it takes longer to make changes. Thus, in the eighties problems in attempting to operate a non-national system of Federal orders in a near-national market and increasing difficulty of making changes in a timely fashion within the framework of a voluntary program for individual markets will grow in importance.

The alternatives in possible coverage of the Federal order system have two dimensions—the grades of milk to be covered and the geographic area. In terms of grade, the options include:

o All milk, both fluid grade and manufacturing grade.
o All fluid grade milk.
o Only enough fluid grade milk to meet fluid product needs plus a reserve.
o Enough fluid grade milk to meet needs for fluid and soft products (ice cream, cottage cheese, yogurt) plus an adequate reserve.

To date, Federal orders have covered only fluid grade milk. Until the sixties, order policy was generally to include only enough fluid grade milk to meet fluid product needs plus a reserve, although the needed reserve had a somewhat elastic definition. In the late sixties and particularly in the seventies, Federal order policy shifted. In general, it has implicitly accepted the proposition that the conversion from Grade B to Grade A was inevitable (see Graf and Jacobson, 1973) and desirable and it set out to accommodate that conversion. Others have argued that the conversion is not inevitable but is a result of the policy decision to facilitate it (Buxton, 1978).

The coverage options in geographic terms include:

o The entire conterminous United States (48 States).
o The 48 States less California.
o The present coverage.

California, Hawaii, and Alaska are sufficiently separated from the rest of the United States dairy economy so that a Federal order system including all of the country except those three States could achieve most of the same things that one including all 50 States could accomplish, but California's isolation is breaking down with growing production and surpluses. The relationship of the dairy industry in those States to the rest of the country is largely through manufactured dairy products markets and the price support program. Such comprehensive coverage would remove the pockets of privilege in State-regulated and unregulated markets.

More comprehensive coverage of Federal orders would permit the system to do a number of things it cannot now do and to do others differently with different results, but it would quite likely require some change from the present voluntary system through producer vote. The present procedure requiring that two-thirds of the producers who vote must approve gives veto power to one-third-plus-one of those voting. One alternative, if producer consent is required, is to reduce the requirement from two-thirds to a majority vote.

One result of a comprehensive Federal order system covering all milk in the United States would be a lesser need for detailed regulation in some areas. A Federal order is a long and complex document. One of the major reasons for the length and complexity of a Federal order is the requirement to treat producers and handlers in similar circumstances in the same way. No handler is to pay a lower price than his competitors for milk used in the same product, except for reasons spelled out in the order. Similarly, no producer is to receive a lower price than other producers in similar circumstances. If there were not a general feeling that they were receiving equal treatment, Federal orders would not be acceptable. Many of the provisions of Federal orders including plant pooling requirements, allocation rules, the rules dealing with reconstitution, treatment of producer-dealers, pricing of out-of-order milk, and numerous others are in the order to provide such equal treatment. But the application results in friction from those who feel disadvantaged by some particular provision.

A comprehensive Federal order system covering all milk in the United States would eliminate the need for regulations dealing with non-federally regulated milk. If all milk, both Grade A and Grade B, were regulated under Federal orders, there would be no need for detailed plant pooling requirements. On the other hand, more detail in regulations would probably be required to determine who gets what in a comprehensive pooling system including all milk.

A related question revolves around the number and size of market orders. Decreasing isolation of milk markets has brought about mergers and territorial expansion of Federal marketing orders over the past 30 years. From a peak of 85 orders in 1962, the number was reduced to 47 in the late seventies, although the area under regulation increased.

When fluid milk markets were relatively isolated, they were defined in terms of a marketing area where handlers competed in selling fluid milk products. Milk delivered to those handlers became reguated under the order. In today's greatly changed markets, an appropriate policy might be:

o A Federal order that is carrying the surplus for unregulated or State-regulated areas should be expanded to include those areas.
o Orders should be merged when a large Federal order is carrying the surplus for a smaller Federal order.
o Federal orders with overlapping supply areas should be merged.
o A Federal order market pool should be at least as large as the pool operated by the largest cooperative in that market.
o A Federal order market pool should be at least as large as the sales area of the largest handler in that market.

Such a policy would mean a substantial reduction in the number of market orders. All orders except those in isolated areas would become at least regional in scope.

One result of such a policy of large regional orders would be that the need for regulations dealing with other-order milk would be minimized. It would also go a substantial distance toward eliminating pockets of privilege and providing equal treatment of producers and handlers in similar circumstances.

Another aspect of the voluntary nature of Federal orders is that of voting. This involves two emotionally charged issues: (1) the percentage of the vote required to approve an order and (2) bloc voting by cooperatives. Clearly, the rights, burdens, and benefits of minority and majority are involved. The process of obtaining a vote for approval is one of negotiation and consensus-building to achieve the required percentage of the vote. Bloc voting by cooperatives gives them an advantage over unorganized producers in voting, but the requirement for two-thirds approval means that a cooperative with less than that proportion must negotiate a consensus with other organizations or unorganized producers. Any change in voting rules would alter the distribution of power and of benefits and burdens.

The Political Economy of Classified Pricing

The existence of classified pricing is an affront to economists, lawyers, and consumer advocates because, as everyone knows, classified pricing is equated with price discrimination. And price discrimination is illegal under the Robinson-Patman Act, although explicitly permitted under the Agricultural Marketing Agreement Act. So the lawyers know that it is just another special favor for agriculture.

And economists can look at the price discrimination diagram and conclude that consumer surplus is smaller and producer surplus larger with price discrimination than without it.

But all that glitters is not price discrimination. Economic price discrimination is defined as charging different net prices to different buyers for the same product (Waugh, 1964, p. 65). Nominal prices are easy to compare. Net prices are much more difficult, since they require substantial amounts of cost information as well as quoted prices. Comparison of prices for the same product requires that the product be identical in space, time, and form utility or that differences in these utilities be costed out into net prices.

It is argued by Masson et al that any difference in the price for milk from, say, a given tankload which is used for different purposes is conclusive evidence of price discrimination and of excessive market power on the part of the seller.

> "Much milk is sold to processors in many markets on a single delivery schedule, delivered into a single intake. This milk satisfies the economic conditions of being identical in time, form, and space." (Masson et al, 1978, p. 66.)

As to any given tankload of milk, the physical characteristics of the part of the lot used for fluid milk or for ice cream or butter-powder are certainly identical. But the value in economic terms of the milk used for ice cream is significantly less than that used for packaged milk. How can this be?

The economic principles embodied in location theory tell us that in a perfect market the value of milk for fluid use at a city plant is given by the cost of transportation of bulk milk from the outer reaches of the milkshed, but the value of milk for use in ice cream is limited by the cost of transporting ice cream or ice cream mix from plants in the surplus production area. The economic forces represented by these principles are at work with or without the intervention of public regulation.

Thus, milk for use in ice cream can only move to a city plant at a price which does not exceed the cost of alternative ingredients—e.g., ice cream mix—from somewhere else, providing there are no restrictions on movement or requirements that only local milk can be used to manufacture ice cream.

But, in this free-flow market, there will be no milk for use in fluid products if the higher price including transportation is not paid, since it obviously then pays the producer better to sell to a manufacturing plant in the outer reaches of the milkshed.

In terms of demand curves, the demand for raw milk for use in ice cream is zero if priced the same as that used in fluid products, since the entire demand has already been met by ice cream mix from more distant areas.

In such a situation, in a perfectly planned world, precisely as much milk would be produced in the local milkshed as was needed for fluid milk products and all ice cream would be made from shipped-in mix. But since cows produce varying amounts, storms interrupt assembly and hauling, and consumers don't behave according to plans, some mismatch between production on the farms and plant use for packaged milk is inevitable. It is obviously less wasteful to use surplus milk (the reserve against running out) for manufactured products than to put it down the drain...which is now illegal in most jurisdictions because it creates additional load at the sewage plant.

Thus, that which is so crystal clear when one looks at a single, isolated tankload of milk takes on very different dimensions when one looks instead at the flow of milk day after day. Because it is the flow of milk that must be priced rather than a single lot, the dimensions of space, time, and form utility are changed. The "market" for that isolated tankload of milk at the dock of the city milk plant is only an incident in an interconnected set of markets for milk in all its forms and for all its products.

Classified pricing of milk was invented to deal with these problems which are basically unchanged in a hundred years. The additional possibility of price discrimination was discovered much later. That had to await the development of market power. 1/

The original term for what is now called classified pricing was "use pricing." As an aid to understanding the problem and the system developed to deal with it, there is much to commend the older term.

1/ The economic concept of price discrimination was much slower to come into use. As far as I can tell, the term was first applied to milk by John Cassells in 1935, building on Joan Robinson.

On the demand side, the value of bulk milk for use in manufactured products is limited by the cost of competing ingredients...the ice cream mix in the preceeding example. But the value of bulk milk for use in packaged fluid milk products is limited only by the competition of bulk milk shipped in from other areas.

On the supply side, there were and are substantial costs of producing and marketing Grade A milk for use in fluid milk products which are not incurred in producing and marketing Grade B milk for use in manufactured products. A supply curve facing the fluid milk plant must reflect all of these costs and, since the price of milk for use in manufactured products ("surplus" milk) is limited by the cost of ingredients, all of the costs are reflected in the Class 1 price.

In the beginning, when sanitary regulations required farmers to meet higher standards in producing milk for sale as fluid milk products, the additional costs of meeting such standards were the major cost factor considered, but not the only one. Estimates of the added production costs over the years have been:

	Cents per hundredweight	Percent of price
1935	46	26.7
1967	25	5.0
1974 Missouri	16	1.9
Upper Midwest	10	1.2

(Black, 1935, p. 118; Hammond and Buxton, 1968; Cummins, 1978.)

In the thirties, fluid milk processors incurred nearly all of the additional costs of marketing bulk milk. The additional costs of operating a receiving and balancing system were incurred largely by processors. Thus, they were not reflected in Class 1 prices since the processor met the costs over and above the prices paid to producers, although they were reflected in wholesale and retail prices of packaged milk. As these functions moved from fluid milk processors to cooperatives in the past 25 years, prices paid by processors to cooperatives for milk used in fluid milk products came to reflect many of these costs. As we have seen, these were reflected largely in over-order payments.

These costs exist and must be covered by the marketing system in one way or another. They are not the result of monopoly power. On the contrary, they would be very much higher if large cooperatives were broken up and the system no longer had the benefits of economies of size.

Costs of Marketing Bulk Fluid Milk. Significant additional costs are involved in marketing bulk milk for use in fluid milk products which are not incurred if the only use is for manufactured products. In a system designed solely for manufactured products, most milk would move directly from the farm to the manufacturing plant and be manufactured into butter, powder, cheese, or other manufactured products with a minimum of delay. The only constraint would be the capacity of the manufacturing plant to handle the volume of milk shipped from farms on a given day. This constraint would only be serious in the flush production season of a year of heavy milk production. Manufactured products are all storable and incongruities between short-term supply and demand are met by varying the volume in storage.

Matching of milk supply and product demand for fluid milk products and soft products cannot be handled through variations in storage, except to a limited extent for ice cream. This gives rise to the necessity for reserves of milk in order to have enough milk available on a given day to meet the needs of fluid milk processors. The costs of supply-demand balancing and coordination include extra hauling costs to move milk to short-supply areas and also to divert reserve supplies to manufacturing and balancing plants, costs of bulk storage used to hold milk supplies to meet peak demand days, personnel and office expenses involved in delivery coordination and rerouting bulk tank trucks, shrinkage resulting from splitting loads and reloading to divert milk to manufacturing plants, general administration attributed to the function of coordinating supplies, health and quality inspection fees on reserve milk, market administration fees on reserve milk, and plant give-up costs. Plant give-up costs are the increase in unit costs which occur when reserve milk is withdrawn from balancing plants to make supplemental shipments of milk to market milk handlers. A study of the inter-mountain region (South Dakota, Colorado, Wyoming, Utah, Idaho, and Nevada) in 1977 found these costs to total 42.3 cents per hundredweight on all milk delivered (table 101). But since the costs of operating a fluid milk marketing system can only be recovered from fluid milk products, the cost per hundredweight of Class 1 milk was over 50 cents. With substantial increases in costs since 1977, costs in that area are now over 60 cents per hundredweight of Class 1 milk (Christensen et al, 1979; see also Babb, 1980).

The inclusion of give-up costs concerns many. One could regard those costs as the price paid for being in the pool with an obligation to ship milk for fluid use. But the costs exist—a balancing plant has higher costs of manufacturing than a pure manufacturing plant because of greater irregularities in supply. In other words, if all milk went directly to manufacturing uses, these costs would be avoided. Since that is the comparison being made, they are a part of the added costs of supplying the fluid milk needs of the market. (See Lasley and Sleight, 1979, for the effects of varying volume on manufacturing costs. See Ling, 1982, for give-up costs on milk for manufacturing use.)

The costs of supply-demand balancing and coordination vary from market to market depending upon the size of the market, the degree of centralization of coordination within a market, and the size of the reserves, among other things. With centralized coordination and large reserves, costs are significantly lower. Centralized coordination can reduce costs by nearly two-thirds, as compared with a system where every plant handles its own reserves (Lasley and Sleight, 1979).

Alternatives. Any pricing system for milk meeting the sanitary requirements for fluid use must recognize and deal with the basic fact that milk which is indistinguishable at the farm is no longer

Table 101--Total cost of coordinating and balancing supply with
demand for 55 fluid milk handlers by seven dairy
cooperatives in the Intermountain area, by type
of plant, 1977

Function	Type of Plant Served			
	Full supply	Major supply	Minor supply	Total
	- - Cents per cwt. of milk delivered - -			
Give-up cost	18.9	62.2	108.7	26.0
Other costs:				
Hauling	10.1	10.1	10.1	10.1
Bulk storage	1.6	1.6	1.6	1.6
Delivery coordination	2.2	2.2	2.2	2.2
Shrinkage	.3	.3	.3	.3
General administration	.3	.3	.3	.3
Health inspection	.8	.8	.8	.8
Market administration	1.0	1.0	1.0	1.0
Total, other costs	16.3	16.3	16.3	16.3
Total, all costs	35.2	78.5	125.0	42.3

Source: Christensen et al, 1979, p. 41.

the same, in an economic sense, when it reaches the fluid milk plant
rather than the manufactured products plant.

Beyond the farm level, there are significant costs in supplying
milk to fluid milk processors that are not incurred when supplying
milk to manufacturers of butter, cheese, and nonfat dry milk. Transportation, the most obvious and significant of these costs, appears
as a separate item under Federal milk marketing orders and is partially reflected in Class 1 and blend prices.

Other costs unique to providing milk to fluid milk processing
plants arise out of the need to provide that milk in the form and on
the schedule desired by the processor. The very significant costs
of providing this kind of service must be covered by the pricing
system, whether or not it takes the form of the present classified
pricing system.

Thus, a classified pricing system or something akin to it is
necessitated by the economics of the situation. The additional costs
of producing and marketing milk for fluid products as compared to
milk for manufactured products must be covered by the pricing system,
either in Federal order minimum class prices or in service charges
of cooperatives. But, over and above these costs, policy options

remain as to how much price discrimination is reflected in Class 1 prices, if any. (See Fallert and Buxton, 1978.)

One option is to reduce Class 1 differentials to minimal levels. This would mean setting the Class 1 differential at the base markets to cover the additional costs of producing and marketing milk for fluid use (or a part of the marketing costs, leaving the remainder to be covered by cooperative service charges) and establishing a geographic structure of Class 1 milk prices related to transportation costs on the amount of milk that needed to move between markets. This would provide a set of minimum Class 1 prices that included no economic price discrimination.

Another option would be to raise Class 1 differentials and, at the same time, lower manufacturing milk price supports by an equal amount. Because of differences in demand elasticities for fluid milk and manufactured dairy products, this would raise total returns to dairy farmers. Returns to Grade A farmers would increase and those to manufacturing milk producers would decrease, unless all milk were pooled. Then, total returns could be pooled and allocated to different groups of producers in various ways.

A third option would be to lower the base (support) price to get closer to a market-clearing level and leave Class 1 differentials at the present levels. This would let competition establish the effective Class 1 prices thru over-order payments. With a closer balance between production and consumption, the overall level of over-order payments would tend to reflect the actual level of transportation charges on needed movements of milk. Maintaining order Class I differentials would provide a "safety net" for unusual circumstances.

Structure of Class 1 Prices

Class 1 differentials can be varied from the existing level, but there are limits on the lower end. The additional costs of producing and handling milk for fluid products, over and above the costs for milk used solely in manufactured products, need to be covered either by Class 1 differentials or by over-order payments to cooperatives.

The relatively inflexible geographic structure of Class 1 prices played a role during the sixties in encouraging milk production in many areas beyond the needs of the fluid milk market. However, changed conditions in 1973 eliminated the surpluses. At the same time, transportation costs for bulk milk increased sharply with escalating fuel prices and general inflation. The Federal order minimum Class 1 price structure has not reflected these increased transportation costs. Cooperatives in most but not all areas adjusted their prices to reflect some portion of those increased costs, coming closer to achieving a Wisconsin-plus-transportation price surface.

But, in 1982, production is rising faster than demand in most parts of the country and the geographic structure of prices should become flatter again. Thus, flexibility is required. The policy question for the longer run is: should that kind of flexibility be provided by changing the Federal order Class 1 price structure as supply-demand conditions change or is it preferable to set the Federal order minimums at levels appropriate for ample supply conditions (when there are surpluses most everywhere) and let market forces create a geographic price structure appropriate for tight supply conditions when they arise? These questions are related to that of minimal prices.

Minimal Prices. The question of how minimal the minimum prices should be applies to policy decisions with respect to the size of the Class 1 differential, the geographic structure of Class 1 prices, payment for marketwide services, and probably other questions.

The nature of the policy decision has been indicated previously with respect to the Class 1 price structure. The appropriate geographic structure of Class 1 prices, as well as the overall level, changes with supply and demand conditions. Present practice changes the level, through changes in the M-W price, but not the structure. The policy choice is between (1) specifying minimum Class 1 prices and letting the market raise prices above that level, creating a new price structure when economic conditions dictate, and (2) adjusting the minimum Class 1 price structure in Federal orders to reflect changing economic conditions at least annually.

If the only options considered are the existing base-zone-plus-transportation structure using 1968 transportation costs or a similar price structure changed only by using current milk transfer costs, the decision comes fairly easily to maintain the existing structure (see Babb and Litzenberg, 1980, pp. 12-13). The current price structure provides some stability without creating incentives for uneconomic movement of milk and would minimize the shock of deregulation, if it were to come. However, if the additional option of a price structure changed annually to reflect supply-demand conditions is introduced, the conclusion is less clearcut. The minimal features of the current structure could be retained to the extent desired by reflecting something less than full transfer costs, if desired, or much the same result could be obtained by assuming a reserve requirement at a minimal level.

Specifying only minimal levels greatly simplifies the task of regulation. It allows market forces to operate more freely. It avoids the situation that, with annual adjustment, prices would be lowered by the Department at times, engendering a very negative reaction by producers. But, when prevailing prices are well above the minimums, it also requires the Department to undertake much more difficult analytical tasks in appraising cooperative milk prices to determine whether or not undue price enhancement may exist.

If the Class 1 price structure specified in Federal orders is changed, say annually, to reflect changing supply and demand conditions, the task of Federal order regulation is made significantly more complex and more sensitive to error. A system of regional supply-demand adjusters could be used to achieve such an annually (or semi-annually) adjusted price structure.

In either case, the base zone is substantially larger than that embodied in present practice. It extends from Minnesota through Wisconsin to New York and Pennsylvania (see Hallberg et al, 1978).

The existence of a structure of Federal order Class 1 prices that conformed with supply and demand conditions would make it possible to use those prices as the basis for analysis of potential undue price enhancement. In such a case, undue price enhancement could be identified provisionally as prices which exceeded Federal minimums plus the cost of services provided by cooperatives. A legal finding of undue price discrimination would still rest on a finding of monopolization or restraint of trade. (See Manchester, 1982).

An integral part of the policy question is: to what extent does public policy wish to encourage, through Federal orders or other means, the development of market power by cooperatives? As has been indicated, a number of dairy marketing cooperatives now have a measure of market power and some price enhancement results. (See Dobson and Salathe, 1979, p. 218.)

However, a change in the basic objective of Federal dairy price policy from stability to substantial income enhancement would complicate the problem of identifying possible undue price enhancement. The kind of economic standard applied in the 1975-76 case (Capper-Volstead Committee, 1976) would no longer be available. Under those conditions, the only standard for identifying the existence of possible undue price enhancement would be cost of service.

Methods of Determining Class Prices

Since the mid-sixties, Class 1 prices have moved up and down with changes in the average price paid for manufacturing grade milk in Minnesota and Wisconsin. Good measures of manufacturing milk prices have been relatively easy to obtain and have provided a sensitive measure of changes in the overall supply-demand balance in the dairy economy.

This system worked well in the sixties and early seventies in providing a mover for the entire class price structure. Events in 1973-75 raised doubts as to its adequacy. Rapid fluctuations in the M-W price reflected short-run supply and demand conditions for all milk, including import actions, but sent misleading signals to producers, handlers, and consumers as to what to expect.

At a time when feed prices were jumping, milk prices dropped. Within a few months, the situation turned around and milk prices rose nearly as sharply as they had dropped. This is undesirable in a system intended to provide some measure of stability. It serves no visible public purpose.

This suggests that a more effective system might be developed which would cut the direct tie that binds Class 1 prices by fixed differentials to the M-W price for manufacturing grade milk. Alternative methods of determining Class 1 prices ("economic formulas") might provide a basis for a better-performing system if supply-demand principles were respected (see Smith and Knutson, 1979).

On the other hand, if 1973-75 is a once-in-a-lifetime occurrence, a change to economic indexes may represent overkill.

A shift to an economic formula for Class 1 prices would mean a major break with the close link between Class 1 and manufacturing milk prices that is a major feature of the existing price structure. It could lead to a situation where Class 1 prices were rising when manufacturing milk prices and perhaps support levels were declining, although such a situation is unlikely with a formula which respects supply-demand principles and could be prevented by the use of appropriate snubbers.

A change in price movers would become necessary if the competitive market for manufacturing grade milk disappeared due either to conversion to Grade A or to extension of Federal order regulation to all milk. The principal alternative to the M-W price for manufacturing grade milk in such circumstances is a product-price formula. It would be based on the plant selling prices of the principal manufactured products—butter, powder, and cheese—with appropriate yield factors and margins to cover processing costs. Such plant selling prices would have to be obtained for butter and cheese—they are now collected only for nonfat dry milk. An annual survey to obtain information on yield factors and manufacturing costs would also be required. (See Jacobsen et al, 1978; Leathers and Hammond, 1980.)

Federal order prices for Class 3 milk—that used in butter, nonfat dry milk, and cheese—have generally been set at the M-W price level since the sixties. It has been proposed that Class 3 Federal order prices be set at the M-W price or the support price, whichever is higher. This would be effective only when the M-W price dips below the support price. This would raise the costs of plants manufacturing butter, powder, and cheese that are regulated under Federal orders including balancing plants whose costs are already higher than those not so regulated and make them noncompetitive in the product markets. They would be subject to a cost-price squeeze that was not affecting their unregulated competitors. It would make it essential that "make allowances" under the support program accurately reflect the current costs of manufacture of butter, powder, and cheese, as the State of California now does.

Distribution of the Proceeds

What prices are paid to different groups of producers? This involves questions both of pricing methods and of pooling methods, including who is pooled (i.e., what coverage of Federal orders). If both fluid grade and manufacturing grade milk producers are covered by Federal orders, they will share in the distribution of the proceeds of Class 1 sales. If only fluid grade milk producers are covered, as at present, manufacturing grade milk producers will not participate. Differing levels of the Class 1 differential, methods of determining class prices, and geographic price structures will yield different levels of prices to producers in various regions, with consequent effects on their incomes.

A related set of options revolves around the question of conversion from Grade B to Grade A production. Present practices encourage conversion, for reasons previously stated, but not very much since order Class 1 differentials now add only 12-21 cents to the blend price of producers in the areas where conversion is occurring. If Class 1 differentials were reduced significantly, there would be less incentive for producers to convert and some producers might revert to Grade B. However, as in the case of farm bulk tanks, a point is reached where it is uneconomical for plants to have dual collection systems and dual intakes and the disappearance of manufacturing grade producers is the result of decisions by plant operators to which the producers must conform or go out of business. In some areas of the upper Midwest, such a point has already been reached. On the other hand, changes in pooling requirements in the the upper Midwest marketing orders (which would mean essentially eliminating them) could make it feasible for nearly all Grade B producers desiring to stay in milk production to convert fairly shortly. This would make all producers eligible for the somewhat higher blend prices under Federal orders, although the differences would narrow further. It would also mean the end of the competitively determined pay price for manufacturing grade milk in Minnesota and Wisconsin and new pricing methods would have to be devised.

Substantial reserves of fluid grade milk are required under the present system, for reasons already discussed. The reserve requirement is on the order of 30 percent of fluid product consumption at the present time, although it varies with market size and organization. Measures that reduce reserve requirements would obviously improve efficiency in milk production and marketing, because more milk would be produced in low-cost areas and subsequently made into manufactured products. A number of actions would help to reduce reserve requirements.

Continued merging of Federal orders would make larger quantities of milk potentially eligible for use at any given plant without whatever additional cost is imposed by allocation or other provisions of the Federal orders. Expanding the Federal order system to include presently nonfederally regulated supplies of milk would also increase somewhat the supplies available for use at any one point.

The most controversial proposal is to remove the disincentives to use reconstituted and filled milk under Federal milk marketing orders. If provisions requiring a handler making reconstituted or filled milk to pay more than the equivalent of Class 1 prices were removed or, alternatively, changed in all orders to the same level as Class 1 prices in the Upper Midwest, these products would be available in markets at a considerable distance from the upper Midwest and could be used to fill the gap between locally available supplies and demand without importing fluid milk from surplus areas. This would only apply in areas at a considerable distance from the surplus production area, since the cost of removing the water, transporting the products, and putting the water back in would exceed the cost of fluid milk imported from the surplus area in areas closer at hand. How much effect it would have on the need for reserve supplies of milk in distant markets is somewhat problematical. Consumers are much slower to switch from one product to another than are manufacturers. It would probably take substantial price differentials to persuade them to switch when reserve supplies were short. This might well mean larger variations in retail prices than consumers are willing to accept. It would certainly not represent stability, as long as consumers did not regard whole milk and reconstituted milk as very close substitutes.

Thus, reconstituted milk would only serve as a reserve if it could be easily substituted in the marketplace for fresh fluid milk and this would only be the case if reconstituted milk was indistinguishable from fresh milk or equally acceptable.

As long as consumers regard reconstituted and fresh milk as different products, reconstituted milk is not a reserve for fresh milk. If a significant group of consumers switched from fresh milk to reconstituted milk and then did not switch back when fresh milk supplies recovered, fluid milk needs in the market would be permanently reduced. When local supplies were reduced to the new level, the one-way shift to reconstituted milk would recur in the short-supply season.

Incentives to Efficiency

One nontrivial objective of milk orders is to provide incentives to efficiency in marketing bulk milk. A principal portion of the problem is to minimize movement of milk. The importance of this has increased with the more than doubling of milk hauling costs in the last ten years as fuel costs soared.

Some milk has been moved to fluid milk plants for the sole purpose of qualifying a country plant to participate in the pool, with equivalent quantities of milk moving in the other direction. This is less prevalent than it once was, due to changes in the regulations, especially the rule which netted out such movements between plants in the same "system."

Shipping requirements have been used to assure that country plants participating in the pool actually are supplying milk for fluid use. Thus, they require extra movement of milk at times when milk is available closer to the market. Similar results can occur when a handler has his own supply from independent producers, if they are distant from the city.

Possible means of providing better incentives to minimize hauling include shipping requirements that vary with distance from the central city and intramarket transportation allowances that provide more incentive to handlers to use close-in milk. There are obvious tradeoffs here between the efficiency that a monopoly cooperative could provide and the benefits of competition.

Protein Pricing and Standards

Changing the standards of composition for skim and lowfat milk to require higher levels of solids-not-fat would increase the demand for nonfat solids and make lowfat and skim milk more palatable to consumers who dislike the "watery" flavor. Lowfat milk with added nonfat solids has been available to consumers for many years at prices which more than covered the added costs of fortification. Sales of fortified lowfat and skim milk have dropped from 76 percent of all lowfat and skim milk in 1969 to 20 percent in 1981. Many consumers appear to have become accustomed to the less-rich flavor of the unfortified product and to prefer it at prevailing price differences. Standards for fluid milk products are established by the individual States and by the Food and Drug Administration. Changing standards in over 50 jurisdictions is a time-consuming process. Raising the minimum for nonfat solids would be a reversal of the downtrend of 20 years.

When standards for other foods have been changed to require fortification, it has been done on the basis of a need for increased amounts of a particular nutrient in the American diet. This argument could not be made effectively for protein. In 1980, the average diet provided more than twice as much protein as needed. The argument appears more persuasive for calcium, as in 1980 the average diet provided only 5 percent more calcium than needed. The Household Food Consumption Survey of 1965-66 found that 31 percent of the households consumed food falling below the recommended dietary allowances for calcium. But it is not milk drinkers who need more calcium. It is those who do not drink milk or only small quantities. So putting more nonfat solids (and thus calcium) into milk products will not help to meet the calcium deficiency of people who do not drink milk.

Until World War II, butterfat carried most of the market value of milk. With the development of the milk drying business during and since World War II, the nonfat solids portion acquired increased value. Early pricing plans provided for a basic price of milk with a variation depending on the butterfat content. This type of pricing plan has generally been used, both in fluid milk and manufacturing milk markets, from the twenties to date. In recent years, plans assigning a specific value to solids-not-fat or protein have been adopted by the State of California and by some cooperatives in other areas.

Component pricing, in which milk is paid for on the basis of solids-not-fat or protein as well as butterfat, could provide incentives to producers to stimulate higher levels of production of nonfat solids or protein. This could add to the surplus of nonfat dry milk by encouraging producers to breed and select cows, feed, and manage on the basis of protein production as well as total milk production. Without raising the standards, component pricing would provide an incentive to fluid milk processors to reduce the level of nonfat solids to avoid the added costs.

Costs of testing for both nonfat solids and butterfat in California are only about 14 percent higher than for butterfat testing alone. An additional pound of nonfat solids in a hundredweight of producer milk was worth from 10 to 39 cents to plant operators in additional yields of nonfat dry milk (10 cents) or American cheese (39 cents) at mid-1982 prices. In the case of fluid whole milk, however, the value of additional solids or more protein is less clear as there is no clear-cut indication that consumers are willing to pay more for additional solids or protein. The increased attention being given to the testing and accounting of milk on a multiple component basis would facilitate transition to a new pricing system.

Multiple component pricing can be introduced in Federal orders whenever there is sufficient interest on the part of producers. With the increased movement of milk between markets and the need to maintain intermarket price alignment, for any new method of pricing to succeed, a uniform plan would have to be acceptable to producers in all markets. A multiple component pricing plan applied to producers also would have to be applied to handlers.

Payments for Marketwide Services

Many cooperatives perform services which benefit not only their own members but also other producers in the market, by performing necessary functions for the market as a whole. But, the cooperatives bear the costs without being able to share them with all those who benefit. These functions include managing and routing the milk supply and providing for disposition of milk not used for fluid products. In principle, it would be possible to reimburse cooperatives for the costs of such market-wide services. However, development of a detailed plan to do so is extremely complex. Very careful and extensive analysis would be required to determine whether or not it is feasible.

If a workable system could be developed to pay cooperatives for market-wide services, it would help to distinguish the cost of services from other charges of cooperatives. It would also come close to solving the free rider problem where nonmembers and members of limited-service cooperatives obtain the advantages of the market-wide services performed by cooperatives but do not pay any of the costs. On the other hand, it would diminish the role of the market in determining effective prices.

Competition and Market Power

Policy regarding competition and market power is less specifically oriented to the dairy industry than the other forms of policy. In general, it is a question of how broad antitrust policy is applied in specific cases in the dairy industry. Occasionally, as when the FTC proclaimed a set of specific merger guidelines for the fluid milk and ice cream processing industries, there is something which might be described as a dairy industry policy. Because of the special features of the dairy industry and of the public role in it, a number of antitrust questions are essentially unique to the dairy industry, for example the question of whether over-order charges by cooperatives represent undue price enhancement.

The effective antitrust policy for the overall economy in the United States is now one of workable competition, as it has been for many years regardless of the rhetoric frequently heard from antitrust officials. The concept of workable or effective competition is based on a rejection of the ideal of perfect competition as the end toward which to aspire. This rejection is based both on the impossibility of achieving perfect competition or even of moving very far in that direction and on a preference for the imperfectly competitive mixed economy that we have, largely because of its dynamic quality. As a policy stance, it says that what we have, warts and all, is preferable to the unattainable ideal of textbook perfect competition because it works. That is not to say it cannot work better.

While it is probably true that the majority of antitrust practitioners would prefer an economy closer to an atomistic ideal, it is evident that their activities since 1890 have not prevented the growth of large firms in the American economy, although they have had some effect in channeling those efforts. For example, the Celler-Kefauver Act of 1950 was widely heralded at the time as preventing the growth in size of firms in the United States. This it clearly has not done, but its application by the antitrust agencies has channeled growth into conglomeration. It has restrained increasing concentration within industries by way of mergers, but it has had zero effect on overall firm growth.

The only area of antitrust policy regarding noncooperative firms where there is some uncertainty is that concerning mergers. The 1967 guidelines regarding mergers and acquisitions in the fluid milk processing and ice cream industries effectively prohibited acquisitions of any sizeable company by the major firms. At the same time, it deliberately set out to encourage the growth of "second-tier" firms in the industry. While application was somewhat inconsistent--it encouraged Southland to expand dramatically but discouraging acquisitions by Dean Foods, a smaller firm--the guidelines together with other changes in the industry (which have already been discussed) resulted in a major restructuring of the fluid milk industry by 1979.

The shift in market power from processors to retailers with the adoption of central milk programs appears to have dulled the interest of large dairy firms in fluid milk processing. The interest of pure conglomerates in fluid milk processing appears to have run its course, as it has in all processing of basic foods. No such acquisition has occurred in the last few years and a resumption of interest by other similar firms seems unlikely. It appears likely that cooperatives will increasingly become the acquirers of large local firms that come on the market, in many cases because the family owners have reached the retirement age.

Policy regarding cooperatives is another matter. The antitrust agencies have long been troubled by cooperatives as a partial exemption to the general sway of the antitrust laws. The views of the staffs of the antitrust agencies found expression in their reports and that of the National Commission for the Review of Antitrust Laws and Procedures in the seventies (U.S. Dept. of Justice, 1977; Lipson and Batterton, 1975; National Comm. for the Review of Antitrust Laws and Procedures, 1979). In that view, cooperatives should have only the right to organize and exist. All other activities would be subject to the antitrust laws, as with other organizations. Membership would be entirely voluntary with no restraints on entering or leaving. Mergers between cooperatives would be subject to the same per se prohibition of horizontal mergers which applies to all other corporations of any size. Two cooperatives could not discuss prices without being accused of price-fixing. Market share would be regarded as a measure of market power without consideration of differences between cooperatives and other organizations. If this image where to prevail, cooperatives would have no market power and would have no means of acquiring any.

An alternative policy toward cooperative market power which would recognize the realities of general antitrust policy would be based on three premises. First, that agricultural marketing cooperatives should be permitted to participate fully in the modern economy --i.e., cooperatives should not be subject to greater limitations on their activities than other firms, after allowance is made for differences between cooperatives and other firms. The qualification is of crucial importance--the same treatment for two persons or firms who are unequal does not yield equitable results.

Second, the partial exemption provided marketing cooperatives under the Capper-Volstead Act recognizes real differences not just between unorganized producers and the noncooperative buyers whom they face but also between the bases for market power of cooperatives and buying firms.

Third, public policy is to allow cooperatives to acquire market power subject to restraints on predatory acts and on undue price enhancement under Section 2 of the Capper-Volstead Act resulting from acts which constitute monopolization or restraint of trade. A possible realistic policy embodying these principles have been sketched elsewhere (Manchester, 1982).

15
Conclusions: Considerations for a Prudent Policymaker

> For every complex human problem, there is a solution
> which is simple, straightforward, and wrong.
>
> --Attributed to Winston Churchill

The careful reader will have discerned by now that the problems to which public involvement in the dairy economy are addressed are indeed complex. Those who already knew the answers before they started reading will have gleaned a few gems here and there that confirm their previously held positions. Hopefully, some of those who started out with simple answers will be sufficiently shaken by the foregoing to consider other possibilities. My own conclusion is that there are no simple answers. This final chapter is a distillation of what seem to me to be considerations that a prudent policymaker would keep in mind as he addressed dairy policy questions in the eighties.

The dairy economy of the seventies and eighties is indeed vastly different from that of the thirties and forties when the principal Federal programs were initiated. But the fundamental differences between production and marketing of milk and of other agricultural products remain unchanged. While the basic legislation dates back to 1937 and 1949, the programs administered under it are also much different from those of the thirties and forties.

The problem of the thirties and forties which captured the most attention has been largely solved--wide disparity between incomes of dairy farmers and those elsewhere in the economy no longer creates a demand for action on equity grounds. With dairy farming now a specialized, commercial operation, it is unlikely to recur as long as a reasonable measure of stability is provided. Thus, while some of the questions have changed, many are quite similar. Milk is still a flow commodity and highly perishable in the raw form and the marketing institutions must deal with those facts. Production adjustments, especially increases in production, are subject to a multi-year biological lag which is unchanged. Production requires large investments that are highly specific to milk output. In the light of those capital investments, farmers' decisions to go into or out of milk production or to make major changes in its scale are not lightly made.

In short, the need for stability in its several dimensions is basically unchanged. At the same time, the system must make possible adjustments in production and marketing with emphasis on orderly change. In a mixed public and private system such as exists in the dairy economy, the results must be perceived to be fair if farmers, processors, distributors, and consumers are to accept them. This means, among other things, that farmers must feel that they are receiving and processors that they are paying prices which are the same as those paid or received by others similarly situated. It also means the level of prices must bear what is regarded as a reasonable relationship to other prices in the economy, including the prices of competitive products and those of the inputs used in production and marketing.

The effect of stability in its various dimensions on the organization of dairy farming is a matter that a prudent policymaker would certainly want to consider. Dairy farming is the last bastion of family farming in animal agriculture in most of the country. The growth of large-scale dairy farming on the west coast and elsewhere indicates that there are alternatives. There is no way of precisely determining the relative contribution of the stability provided by Government programs and other economic and technical factors to the preservation of the family dairy farm, but one could probably get general agreement that there is some relationship. The perceived social and economic benefits of family farming are not insignificant. The policymaker who felt constrained to preserve those benefits would be extremely reluctant to recommend the removal of a minimal stop-loss price support program.

Some degree of stability could be provided by contracting in the absence of public programs, but no one except the Government would be willing to contract with farmers to provide a guaranteed minimum price.

Another aspect of stability to be considered is that the relatively stable grain prices which prevailed from the end of the Korean War until 1972 are unlikely to return. With the U.S. feed grain and oilseed economy increasingly export-oriented, grain prices—the principal cash cost of dairy production—can be expected to respond to world developments and be much less stable than they were. Since the feed price stability of the fifties and sixties was rooted in the excess capacity of the grain-oilseed sector, world market conditions have had a much greater influence on grain and oilseed prices since 1972 and this condition can be expected to persist for an extended period. Thus, dairy production costs will be considerably more variable than they were before 1972. The prudent policymaker would take this into consideration in determining what sort of mover might be used for a dairy price support program.

A choice to use public programs to achieve significant income enhancement for dairy farmers in addition to stability makes major changes:

o Supply control is needed to control overproduction
   and government costs.
o Profitability depends on Government programs, not
   just in production but also in processing.

The domestic casein industry disappeared because nonfat dry milk became more profitable under guarantees of the wartime program and continued to pay processors better under the price support program.

It seems unlikely that the major components of milk—butterfat and nonfat solids—can compete directly with products of vegetable origin in most food ingredient uses. Whey products can compete because the raw material is almost costless to the processor who has nearly recovered the cost of milk from the sale of cheese. Whey can compete with vegetable products but it does so at a considerable cost to nonfat dry milk with which it also competes.

The booming sales of imitation and filled cheese are indicative of a set of developments which the prudent policymaker will ignore at his peril. The growth of the fast food business and of the market for highly processed food products for home use provides opportunities for products such as imitation and filled cheese. In the away-from-home market, the labels on packaged foods disappear before the ultimate consumer ever sees them. In highly processed foods, while the label is there for the consumer to read, the combination of flavors in a pizza, for example, is such that imitation cheese is probably generally indistinguishable from natural cheese. The growth in these two market segments means that the potential for imitation and filled cheese to replace natural cheese is large. Imitation cheese uses vegetable fat and imported casein. Further growth in the volume of imported casein can be checked under existing legislation but the substitution of vegetable fat cannot. It is probably not possible to lower milk prices enough to make natural cheese competitive with filled or imitation cheese and still get the milk produced.

In considering the problems that market orders must deal with, the policymaker would ignore at his peril the continued existence of the free rider problem. The growth in size of milk marketing cooperatives has not meant that the opportunities for free riders have disappeared. On the contrary, the shift of the function of marketing bulk milk (including handling and balancing) from proprietary handlers to full-service cooperatives has increased opportunities for those not performing those services to cut prices and let the firms performing them carry the cost for the entire market. The trick is to put some limits on opportunities for free riding while maintaining some room for maneuver for competitive forces. A regulated pricing system that provided automatic recovery of all costs incurred in handling and balancing the bulk milk market would provide too little incentive for efficient operation and encourage what Harvey Leibenstein calls X-inefficiency. At the other extreme, unfettered competition would make it impossible for any firm—nowadays, any cooperative but also true for proprietary firms—to perform the balancing function for all or most of the market and achieve the substantial economies of size which are possible.

Consideration of classified pricing must start with an understanding of the functions that it performs. Classified pricing is intended to provide an element of stability in prices in the face of cyclical, seasonal, and short-run variations in supply that are quite different from the variations on the demand side. The Class 1 differential provides a means of recovering the substantial additional costs of producing and marketing fluid grade milk. The possibility of price discrimination—charging different prices for the same economic good—only enters after those costs are covered. It is easy to forget that those costs can only be recovered on milk sold for use in fluid products, since fluid grade milk used in manufactured products must sell at prices which are competitive with manufacturing grade milk. Equating all of the Class 1 differential with price discrimination is the grossest of errors.

In considering the question of where milk should be produced, the prudent policymaker would keep in mind the economic principles involved. The quick answer often is that, in an economically rational world, milk should be produced where production costs are the lowest. But that ignores the not-insignificant costs of transporting milk which are sharply higher now than ten years ago. The correct principle is that milk should be produced where the combined costs of production, transportation, and marketing are the lowest. If it were possible to produce all the milk for the entire country in the upper Midwest without increasing costs there, an impossible assumption, milk prices would still be higher in other parts of the country by the cost of moving milk there. In some areas, they would be higher than they are now.

Government policy for a century has been to permit the development of farmer cooperatives with the twin objectives of offsetting the market power of buyers of agricultural products and promoting efficency in their distribution. Since the twenties, policy has been to encourage, rather than merely to permit, cooperative development. At the same time, limits were placed on the exercise of market power by cooperatives by Section 2 of the Capper-Volstead Act of 1922 which forbids cooperatives from unduly enhancing prices. This policy was reaffirmed by the Congress as recently as 1980 in the FTC Authorization Act of that year. In considering appropriate policy towards cooperative market power in the eighties, the policymaker would give appropriate weight to the considerations discussed in Chapter 14, including effective general antitrust policy and the differences between cooperatives and other business organizations.

Orderly marketing is clearly desirable as an abstraction—one would hardly wish to be thought to favor disorderly marketing any more than original sin. But an excess of government-imposed order becomes rigidity that stifles markets and innovation in products and marketing arrangements. The prudent policymaker would endeavor to supply a modicum of order to offset the inherent instability of milk markets.

References

Agnew, Donald B. (1957)
How Bulk Assembly Changes Milk Marketing Costs, U.S. Agric. Mktg.
Serv., Mktg. Res. Rpt. No. 190, July, 91 pp.

Alagia, D. Paul, Jr. (1979)
"Federal Milk Market Order Regulations: A Defense, *"South Dakota
Law Review,* Vol. 24, Summer.

American Dry Milk Institute
Census of Dry Milk Distribution and Production Trends, annual.

Babb, Emerson M. (1963)
Intermarket Milk Price Relationships, Purdue Univ. Agric. Expt.
Sta., Res. Bull. No. 760, Jan., 23 pp.

_____, D.E. Banker, and G.L. Nelson (1976)
Price Relationships Among Federal Milk Marketing Orders, Purdue
Univ. Agric. Expt. Sta., Sta. Bul. 146, Nov., 32 pp.

_____, D.A. Bessler, and J.W. Pheasant (1979)
Analysis of Over-Order Payments in Federal Milk Marketing Orders,
Purdue Univ. Ag. Expt. Sta., Sta. Bull. No. 235, Aug., 103 pp.,
and update by Professor Babb.

_____ (1980)
*Milk Marketing Services Provided by Cooperative and Proprietary
Firms,* Purdue Univ. Agric. Expt. Sta., Sta. Bull. No. 279, May,
33 pp.

_____ and K.K. Litzenberg (1980)
*Restructuring Class I Price Differentials--Theoretical Considera-
tions,* paper at Southwest Milk Marketing Conf., April 22, 14 pp.

_____ and Robert D. Boynton (1981)
"The Accomplishment of Dairy Program Objectives," in *1981 Agricultural Outlook,* Jan., pp. 279-285.

Baker, Burton A., and Rudolph K. Froker (1945)
The Evaporated Milk Industry Under Federal Marketing Agreements,
Wis. Agric. Expt. Sta., Res. Bull. 156, Sept., 91 pp.

Bartlett, Roland W. (1961)
*An Analysis of Specific Comparisons and Conclusions in House Report
No. 231 on "Small Business Problems in the Dairy Industry,"*
IL Agric. Expt. Sta., Res. Rpt. AER-46, Sept., 7 pp.

Bateman, Fred (1968)
"Improvment in American Dairy Farming, 1850-1910: A Quantitative
Analysis," *Jnl. of Econ. Hist.,* 24:255-273, June.

Beal, George Max (1944)
*Economic Factors Affecting the Production of Fluid Milk in the Los
Angeles County Marketing Area,* CA State Depart. of Agric., March.

_____ and Henry H. Bakken (1956)
Fluid Milk Marketing, Mimir Publishers, Inc., 556 pp.

Black, John D. (1935)
The Dairy Industry and the AAA, The Brookings Institution, 520 pp.

Bormuth, W.D., and D.B. Jones (1959)
Plants Manufacturing Dairy Products by Production-Size Groups and Geographic Distribution, 1957, U.S. Agric. Mktg. Serv., AMS-301, March, 20 p.

_____ and T.L. Cryer (1965)
Plants Manufacturing Dairy Products...by Production-Size Groups and Geographic Distribution, 1963, U.S. Stat. Rpt. Serv., SRS-5, March, 20 pp.

Boulding, Kenneth E. (1980)
"The Moral Environment of Public Policy," in *Increasing Understanding of Public Policies and Programs--1980,* Farm Foundation, pp. 3-13.

Boynton, Robert D., and Emerson M. Babb (1982)
Grade A Dairy Farmers' Perceptions of Milk Buyer Performance: The Findings of a National Survey, Purdue Sta. Bull. No. 367, 32 pp., March.

Bressler, Raymond G., Jr. (1942)
Economies of Scale in the Operation of Country Milk Plants, New England Research Council on Marketing and Food Supply in cooperation with the New England Agric. Expt. Stations and the U.S. Dept. of Agric. June, 92 pp.

_____ (1952)
City Milk Distribution, Harvard Econ. Studies, Vol. XCI, Harvard Univ. Press, 398 pp.

_____ (1958)
"Pricing Raw Product in Complex Milk Markets," *Agric. Econ. Res.,* 10:113-130, Oct.

_____ and Richard A. King (1978)
Markets, Prices, and Interregional Trade, 426 pp.

Butz, William T. (1962)
Geographic Structure of Milk Prices, 1960-61, U.S. Econ. Res. Serv., ERS-71, Aug., 11 pp.

_____, et al (1968)
The Impact of Filled and Nondairy Products, Milk Ind. Found, May, 63 pp.

Buxton, Boyd M., and Jerome W. Hammond (1974)
"Social Cost of Alternative Dairy Price Support Levels," *Am. Jnl. Agric. Econ.*, 56: 289-291, May.

_____ (1977)
"Welfare Implications of Alternative Classified Pricing Policies for Milk," *Am. Jnl. Agric. Econ.*, 59:525-9, Aug.

_____ (1978)
The Disappearance of the Grade B Milk Market---A Matter of Policy Choice? U.S. Econ., Stat., and Coop. Serv., Agric. Econ. Rpt. No. 416, Dec., 14 pp.

_____ (1979)
"Post-Regulation Research: Milk Marketing Order Regulations," *Am. Jnl. Agric. Econ.*, 61:779-786, Nov.

Cadwallader, Richard C. (1938)
Government and Its Relationships to Price Standards in the Milk Industry, Minnesota Law Review, 22:789-835.

Capper-Volstead Committee (1976)
The Question of Undue Price Enhancement by Milk Cooperatives, U.S. Dept. of Agric., Dec., 75 pp.

Capponi, Silvio (1982)
Over-Order Payments in Federal Milk Order Markets, U.S. Agric. Mktg. Serv., 12 pp.

Carley, Dale H., and T.L. Cryer (1964)
Flexibility of Operation in Dairy Manufacturing Plants: Changes 1944 to 1961, U.S. Econ. Res. Serv. and Stat. Rptg. Serv., Agric. Econ. Rpt. No. 61, Oct., 44 pp.

Cassells, John M. (1935)
"The Fluid-Milk Program of the Agricultural Adjustment Administration," *Jnl. of Political Economy*, 43:482-505, August.

_____ (1937)
A Study of Fluid Milk Prices, Harvard Univ. Press, 203 pp.

Christ, Paul G. (1980)
"An Appraisal of the U.S. Government's Role in Milk Pricing," *Am. Jnl. Agric. Econ.*, 62:279-287, May.

Christensen, Rondo A., Douglas E. Petterson, and A. H. Swainston (1979).
The Function and Cost of Market Milk Reserves and Balancing Supply with Demand, Utah State Univ., Econ. Res. Inst., Study Paper 79-8, Sept., 54 pp.

Cochrane, Willard W. (1980)
"Some Nonconformist Thoughts on Welfare Economics and Commodity Stabilization Policy," *Am. Jnl. Agric. Econ.*, 62:508-511, Aug.

Cook, Hugh L., et al (1978)
*The Dairy Subsectors of American Agriculture: Organization and Ver-
tical Coordination,* NC-117 Monograph 5, N.C. Reg. Pub. 257, Nov.,
157 pp.

Cowden, Joseph M. (1956)
Bulk Milk Handling in 1955, U.S. Farmer Coop. Serv., Gen. Rpt. 22,
April, 38 pp.

_____ and Harry C. Trelogan (1948)
Flexibility of Operation in Dairy Manufacturing Plants, U.S. Dept.
of Agr., Circ. 799, Sept., 40 pp.

Cummins, David E. (1978)
Comparison of Production Costs of Grade A and Grade B Milk, U.S.
Econ., Stat., and Coop. Serv., ESCS-5, Jan.

Dahlgran, Roger A. (1980)
"Welfare Costs and Interregional Income Transfers Due to Regulation
of Dairy Markets," *Am. Jnl. Agric. Econ.*, 62:288-296, May.

Dairy and Food Industries Supply Assoc. in cooperation with National
Assoc. of Dairy Euipment Manufacturers (1966)
1965 Farm Milk Tank Survey, 8 pp.

_____ (1967)
1966 Farm Milk Tank Survey, 8 pp.

Dairy Record, July 3, 1935.

Dobson, W.D., and Boyd M. Buxton (1977)
*Analysis of the Effects of Federal Milk Orders on the Economic Per-
formance of U.S. Milk Markets,* Wis. Agric. Expt. Sta., Res. Bull.
R2897, 36 pp., Oct.

_____ and L.E. Salathe (1979)
"The Effects of Federal Orders on the Economic Performance of U.S.
Milk Markets," *Am. Jnl. Agric. Econ.*, 61:213-27, May.

Douglass, William (1755)
*A Summary Historical and Political of the British Settlements in
North America.*

Elsworth, R.H. (1936)
*Statistics of Farmers' Cooperative Business Organizations, 1920-
1935,* U.S. Farm Credit Admin.

294

Elwood, Robert B., Arthur A. Lewis, and Ronald A. Struble (1941)
Changes in Technology and Labor Requirements in Livestock Production: Dairying, U.S. Bu. of Agric. Econ. in cooperation with Works Projects Admin., National Research Project, WPA, Report No. A-14, June, 86 pp.

Emery, Harlan J., et al (1969)
Dairy Price Support and Related Programs, 1949-1968, U.S. Agric. Stabilization and Conservation Service, Agric. Econ. Rpt. No. 165, July, 144 pp.

Fallert, Richard F., and Boyd M. Buxton (1978)
Alternative Pricing Policies for Class I Milk Under Federal Marketing Orders--Their Economic Impact, U.S. Econ., Stat., and Coop. Serv., Agric. Econ. Rpt. No. 401, May, 65 pp.

_____ (1971)
A Survey of Central Milk Programs in Midwestern Food Chains, U.S. Econ. Res. Serv., Mktg. Res. Rpt. No. 994.

Fones, Roger W., Janet C. Hall, and Robert T. Masson (1977)
Federal Milk Marketing Orders and Price Supports, edited by Paul W. MacAvoy, Am. Enterprise Inst., 168 pp.

Foote, Richard J. (1941)
"Federal Relief and Dairy Products Marketing Association Purchases and Stocks of Dairy Products, 1933-41," U.S. Bu. Agric. Econ., *The Dairy Situation, Dec.,* pp. 8-13.

_____ (1947)
"Wartime Dairy Policies," *Jnl. Farm Econ.,* 29:679-683.

Forker, Olan D., and Richard D. Aplin (1970)
Summary of Resale Milk Price Control Survey, Cornell Univ., 9 pp.

Fortune (1939a)
"Milk in Chicago," *Fortune,* Nov., pp. 80-81, 124, 126, 128.

Fortune (1939b)
"Let 'em Drink Grade A," *Fortune,* Nov., pp. 83-84, 128, 131-136.

Frank, Gary G., G.A. Peterson, and Harlan Hughes (1977)
"Class 1 Differentials: Cost of Production Justification," *Econ. Issues,* No. 8, Dept. of Agric. Econ., Univ. of Wis., April.

Freemyer, Glenn W. (1950)
History and Analysis of Milk Supply Problems in the St. Louis Market, U.S. Prod. and Mktg. Admin., Oct., 156 pp.

French, Charles E., and T.C. Walz (1957)
"Impact of Technological Developments on the Supply and Utilization of Milk," *Jnl. Farm Econ.*, 39:1159-70, Dec.

Friedmann, Karen (1973)
"Victualling Colonial Boston," *Agric. History*, 47:189-205, July.

Froker, R.K., A.W. Colebank, and A.C. Hoffman (1939)
Large-Scale Organization in the Dairy Industry, U.S. Bu. of Agric. Econ., Circ. No. 527, July, 67 pp.

Fuchs, A.W., and L.C. Frank (1938)
Milk Supplies and Their Control in American Urban Communities of Over 1,000 Population in 1936, U.S. Public Health Serv, Pub. Health Bull. No. 245, Dec., 70 pp.

Gates, Paul W. (1960)
The Farmer's Age: Agriculture, 1815-60, White Plains: M.E. Sharpe, Inc., 460 pp.

Gaumnitz, E.W., and O.M. Reed (1937)
Some Problems in Establishing Milk Prices, U.S Agric. Adj. Admin., Sept., 227 pp.

Gessner, Anne L. (1959)
Integrated Dairy Operations Through Farmer Cooperatives, U.S. Farmer Coop. Serv., Gen. Rpt. No. 69, Nov., 39 pp.

Gilbert, A.W. (1919)
"Work of the Federal Milk Commission" [for New England], in *Proceedings of the 87th annual Meeting of the New York State Agric. Society*, N.Y. State Dept. of Farms and Markets Bull. 124, pp. 68-81.

Godwin, Marshall, chairman (1975)
Price Impacts of Federal Market Order Programs. Report of the Interagency Task Force, U.S. Farmer Coop. Serv., Spec. Rpt. 12, Jan. 7, 58 pp.

Gold, Norman L., A.C. Hoffman, and Fred V. Waugh (1940)
Economic Analysis of the Food Stamp Plan, U.S. Dept. Agric., Spec. Rpt., 98 pp.

Gordon, Robert M., and Steve H. Hanke (1978)
"Federal Milk Marketing Orders: A Policy in Need of Analysis," *Policy Analysis*, Winter, pp. 23-31.

Graf, Truman F., and Robert E. Jacobson (1973)
Resolving Grade B Conversion and Low Class I Utilization, Pricing and Pooling Problems, Wis. Agric. Expt. Sta., Res. Bull. R2503, June, 77 pp.

Greco, Anthony J. (1982)
State Regulation of Fluid Milk and the Processor-Retailer Margin,
paper presented at Southern Agric. Econ. Assoc., Feb. 8, 29 pp.

Gresham, Mary, and Joseph W. Block (1942)
Labor Aspects of the Chicago Milk Industry, U.S. Bu. of Labor
Stat., Bull. No. 715, June, 53 pp.

Groves, Francis W., and Truman F. Graf (1965)
An Economic Analysis of Whey Utilization and Disposal in Wisconsin,
Univ. of Wis., Agric. Econ. 44, July, 61 pp.

Guth, James L. (1982)
"Farmer Monopolies, Cooperatives, and the Intent of Congress: Ori-
gins of the Capper-Volstead Act," *Agric. History,* 56:67-82, Jan.

_____ (1981)
"Herbert Hoover, the U.S. Food Administration, and the Dairy Industry, 1917-1918," *Business History Review,* 55:170-187, Summer.

Hallberg, Milton C. (1975)
"State Controlled Milk Markets. How Well Do They Perform," *Farm
Economics,* Penn. Coop. Ext. Serv., Nov., 4 pp.

_____, D.E. Hahn, R.W. Stammer, G.J. Elterich, and
C.L. Fife (1978)
*Impact of Alternative Federal Milk Marketing Order Pricing Policies
on the United States Dairy Industry,* Penn. Agric. Expt. Sta. Bull.
818, May, 37 pp.

_____ (1980)
Stability in the U.S. Dairy Industry Without Government Regulations?
Dept. Agric. Econ. and Rural Soc. Staff Paper 37, Penn. State Univ.

_____ (1982)
"Cyclical Instability in the U.S. Dairy Industry without Government
Regulations," *Agric. Econ. Res.,* 34:1-11, Jan.

Hammerberg, Donald O., and William G. Sullivan (1942)
*Efficiency of Milk Marketing in Connecticut. 2. The Transportation
of Milk,* Storrs Agric. Expt. Sta., Bull. 238, Feb., 29 pp.

Hammond, Jerome W. (1967)
Marketing and Pricing of Butter, Minn. Agric. Expt. Sta., Misc. Rpt.
77, March.

_____, and Boyd M. Buxton (1970)
*Consequences of Changing Production Standards for Manufacturing
Grade Milk,* Minn. Agric. Ext. Serv., Spec. Rpt. No. 37, 16 pp.

_____, Boyd M. Buxton, and Cameron J. Thraen (1979)
*Potential Impacts of Reconstituted Milk on Regional Prices, Utiliza-
tion and Production,* Minn. Agric. Expt. Sta., Sta. Bull. 529.

Harris, Edmond S., and Irwin R. Hedges (1948)
Formula Pricing of Milk for Fluid Use, U.S. Farm Credit Admin.,
Misc. Rpt. 127, Dec., 34 pp.

_____ (1958)
Classified Pricing of Milk. Some Theoretical Aspects, U.S. Agric.
Mktg. Serv., Tech. Bull. No. 1184, April, 106 pp.

_____ (1966)
Price Wars in City Milk Markets, U.S. Agric. Mktg. Serv., Agric.
Econ. Rpt. No. 100, Oct., 95 pp.

_____ (1967)
"Impact of Transportation Changes on Price Structures in City Milk
Markets," *Jnl. Farm Econ.*, 49:4:844-851, Nov.

Hayenga, Marvin L. (1979)
Cheese Pricing Systems, NC-117 Working Paper 38, Aug.

Hedlund, Glenn W., chairman (1964)
*Market Structure, Competition and Regulation in the Distribution of
Fluid Milk*, prepared by Committee on Milk Marketing appointed by
Governor Nelson A. Rockefeller, April, 72 pp.

Herrmann, Louis F., and William C. Welden (1941)
Distribution of Milk by Farmers' Cooperative Associations, U.S. Farm
Credit Admin., Circ. C-124, June, 72 pp.

_____ and _____ (1942)
Pre-War Developments in Milk Distribution, U.S. Farm Credit Admin.,
Misc. Rpt. 62, Nov., 24 pp.

_____ (1950)
"Costs of Importing Milk, Memphis, Tennessee, 1948," *The Mktg. and
Trans. Situation*, MTS-80, Jan., pp. 6-15.

_____ and D.B. Agnew (1959)
"Shipments of Milk by Bulk Tank in the United States," *Proceedings
of XV International Dairy Congress*, pp. 2011-2014.

_____ and Helen V. Smith (1959)
"Geographic Structure of Milk Prices, 1957-58," *The Mktg. and Trans.
Situation*, MTS-134, July, pp. 29-32.

Hirsch, Donald E. (1951)
Dairy Co-ops During World War II, U.S. Farm Credit Admin., Spec.
Rpt. 217, Feb., 69 pp.

Hoffman, A.C. (1940)
Large-Scale Organization in the Food Industries, U.S. Temporary National Economic Committee, Investigation of Concentration of Economic Power, Monograph No. 35, 174 pp.

Ippolito, Richard A., and Robert T. Masson (1978)
"The Social Cost of Government Regulation of Milk," *Jnl. of Law and Econ.*, XXI:1:33-65, April.

Jacobson, Robert E., Jerome W. Hammond, and Truman Graf (1978)
Pricing Grade A Milk Used in Manufactured Dairy Products, Ohio Agric. Res. and Dev. Ctr., Res. Bull. 1105, Dec., 70 pp.

_____ (1978)
Possibilities for Correcting Inequities in Super-Pool Pricing in Fluid Milk Markets, paper at Dairy and Food Industry Conf., Columbus, Ohio, Feb 15.

Joint Legislative Committee (1933)
Report of the Joint Legislative Committee to Investigate the Milk Industry in New York, New York Legislative Document 114. Prepared and edited by Leland Spencer.

Jones, W. Webster (1970)
Sanitary Regulation of the Fluid Milk Industry. Inspection, Cost, and Barriers to Market Expansion, U.S. Econ. Res. Serv., Mktg. Res. Rpt. No. 889, June, 45 pp.

_____ (1973)
Milk Sales in Foodstores: Pricing and Other Merchandising Practices, U.S. Econ. Res. Serv., Mktg. Res. Rpt. No. 983, Jan., 28 pp.

_____ (1977)
Economics of Butter Production and Marketing, U.S. Econ. Res. Serv., Agric. Econ. Rpt. No. 365, March, 37 pp.

_____ and Floyd A. Lasley (1980)
Milk Processor Sales, Costs, and Margins, U.S. Econ., Stat., and Coop. Serv., ESCS-77, Jan., 17 pp.

_____ (1981)
Milk Processor-Distributors Sales, Costs and Margins, 1979, U.S. Econ. Res. Serv., ERS-666, Oct., 18 pp.

Kelly, Ernest, and Clarence E. Clement (1931)
Market Milk, John Wiley & Sons, Inc., 489 pp.

Kilmer, Richard L., and David E. Hahn (1978)
Effects of Market Share and Anti-Merger Policies on the Fluid Milk Processing Industry, Ohio Agric. Res. & Dev. Ctr., Res. Bull. 1097, Feb., 14 pp.

Knutson, Ronald D. (1971)
Cooperative Bargaining Developments in the Dairy Industry, 1960-70...With Emphasis on the Central United States, U.S. Farmer Coop. Serv., FCS Res. Rpt. No. 19, August, 36 pp.

_____, chairman (1972)
Milk Pricing Policy and Procedures. Part I. The Milk Pricing Problem, Milk Pricing Advisory Comm., U.S. Dept. of Agric., March, 48 pp.

_____, chairman (1973)
Milk Pricing Policy and Procedures. Part II. Alternative Pricing Procedures, Milk Pricing Advisory Comm., U.S. Dept. of Agric., March, 90 pp.

Krueger, Eugene, and Lillian Z. Woldenberg (1979)
"Cooperatives in Federal Milk Order Markets," *Federal Milk Order Market Statistics for June 1979,* U.S. Agric. Mktg. Serv., FMOS-234, Sept., pp. 36-38.

_____, and Doris Currier (1982)
"Cooperatives in Federal Milk Order Markets," *Federal Milk Order Market Statistics for December 1981,* U.S. Agric. Mktg. Serv., FMOS-264, Feb., pp. 41-45.

_____ (1981)
"An Update of Milk Control Programs," *Dairy Situation,* DS-386, Sept.

Lasley, Floyd A. (1964)
Coordinating Fluid Milk Supplies in the Oklahoma Metropolitan Milk Market, U.S. Econ. Res. Serv., Mktg. Res. Rpt. No. 686, Nov., 56 pp.

_____ (1965)
Geographic Structure of Milk Prices, 1964-65, U.S. Econ. Res. Serv., ERS-258, 7 pp.

_____ (1966)
Coordinating Fluid Milk Supplies in the Pittsburgh Market, U.S. Econ. Res. Serv., Mktg. Res. Rpt. No. 746, March, 52 pp.

_____ (1972)
Subdealers--A Major Form of Milk Distribution in New York State, U.S. Econ. Res. Serv., unpublished report, 30 pp.

_____ (1977)
Geographic Structure of Milk Prices, 1975, U.S. Econ. Res. Serv., Agric. Econ. Rpt. No. 387, Oct., 19 pp.

_____ and Lynn G. Sleight (1979)
Balancing Supply with Demand for Fluid Milk Markets--A Cost Comparison, U.S. Econ., Stat., and Coop. Serv., Oct., 38 pp.

_____, W. Webster Jones, and Leah Sitzman (1979)
Milk Dealers' Sales and Costs: A Trend Analysis, 1952-77, U.S. Econ., Stat., and Coop. Serv., ESCS-62, July, 14 pp.

Leathers, Howard, and Jerome W. Hammond (1980)
An Evaluation of Product Price Measures for Use in Federal Order Pricing, Univ. of Minn., Econ. Rpt. ER80-4, Oct., 47 pp.

_____ (1982)
Dairy Product Prices, Minn. Agric. Expt. Sta., forthcoming.

Lee, James D. (1950)
Dairy Production and Marketing Statistics, Northeastern Dairy Conference, March.

Ling, K. Charles (1982)
Pricing Plans for Managing Seasonal Deliveries by Dairy Cooperatives, U.S. Agric. Coop. Serv., ACS Res. Rpt. No. 22, August, 18 pp.

Lininger, F.F. (1935)
Dairy Products Under the A.A.A., The Brookings Institution, Pamphlet 13, 99 pp.

Lipson, Frank, and Clint Batterton (1975)
Staff Report on Agricultural Cooperatives, U.S. Fed. Trade Comm., Bu. of Competition, Sept.

Lough, Harold W. (1975)
The Cheese Industry, U.S. Econ. Res. Serv., Agric. Econ. Rpt. No. 294, July, 71 pp.

_____ (1980)
Cheese Pricing, U.S. Econ. and Stat. Serv., Agric. Econ. Rpt. No. 462, Dec., 41 pp.

_____ (1981)
Fluid Milk Processing Market Structure, U.S. Econ. & Stat. Serv., ESS Staff Report No. AGESS810415, April, 14 pp., and update.

Manchester, Alden C. (1964)
Nature of Competition in Fluid Milk Markets...Market Organization and Concentration, U.S. Econ. Res. Serv., Agric. Econ. Rpt. No. 67, Feb., 76 pp.

_____ and Leah Sitzman (1964)
Market Shares in Fluid Milk Markets, U.S. Econ. Res. Serv., Rev. April, 73 pp.

_____ (1966)
"Dairy Marketing," in *Agricultural Markets in Change,* U.S. Econ. Res. Serv., Agric. Econ. Rpt. No. 95, July, pp. 150-169.

_____ (1968a)
The Structure of Fluid Milk Markets. Two Decades of Change, U.S. Econ. Res. Serv., Agric. Econ. Rpt. No. 137, July, 51 pp.

_____ (1968b)
Fluid Milk Markets. Number of Handlers and Market Shares, 1950-65, U.S. Econ. Res. Serv., Stat. Bull. No. 428, June, 83 pp.

_____ (1971)
Pricing Milk and Dairy Products. Principles, Practices, and Problems, U.S. Econ. Res. Serv., Agric. Econ. Rpt. No. 207, June, 60 pp.

_____ and Floyd A. Lasley (1972)
"Dairy Products," in *Market Structure of the Food Industries,* U.S. Econ. Res. Serv., Mktg. Res. Rpt. No. 971, Sept., pp. 21-34.

_____ (1973)
Sales of Fluid Milk Products, 1954-1972, U.S. Econ. Res. Serv., Mktg. Res. Rpt. No. 997, June, 11 pp.

_____ (1974a)
Market Structure, Institutions, and Performance in the Fluid Milk Industry, U.S. Econ. Res. Serv., Agric. Econ. Rpt. No. 248, Jan., 40 pp.

_____ (1974b)
Supplement to Agric. Econ. Rpt. No. 248, U.S. Econ. Res. Serv., Aug., 147 pp., and *Corrections and Additions,* 1981.

_____ (1977)
Issues in Milk Pricing and Marketing, U.S. Econ. Res. Serv., Agric. Econ. Rpt. No. 393, Dec., 15 pp.

_____ (1978)
Dairy Price Policy: Setting, Problems, Alternatives, U.S. Econ., Stat., and Coop. Serv., Agric. Econ. Rpt. No. 402, April, 65 pp.

_____ and Richard A. King (1979)
U.S. Food Expenditures, 1954-78. New Measures at Point of Sale and by Type of Purchaser, U.S. Econ., Stat., and Coop. Serv., Agric. Econ. Rpt. No. 431, Aug., 26 pp.

_____ (1982)
The Status of Agricultural Marketing Cooperatives Under Antitrust Law, U.S. Econ. Res. Serv., ERS-673, Feb., 48 pp.

Mann, Jitendar S. (1977)
"Techniques to Measure Social Benefits and Costs in Agriculture: A Survey," *Agric. Econ. Res.,* 29:115-26, Oct.

Market Administrator, New York-New Jersey Milk Marketing Area (1982)
The Market Administrator's Bulletin, Vol. 42, pp. 21-29.

Masson, Robert, Roger Fones, and Janet C. Hall (1978)
Response to the USDA Comments on the Department of Justice Report on Milk Marketing, U.S. Dept. of Justice, Antitrust Div., Oct., 113 pp.

302

_____ and Philip M. Eisenstat (1978)
"The Pricing Policies and Goals of Federal Milk Order Regulations:
Time for Reevaluation," *South Dakota Law Rev.*, 23:662-697, Summer.

_____ and _____ (1980)
"Welfare Impacts of Milk Order and the Antitrust Immunities for
Cooperatives," *Am. Jnl. Agric. Econ.*, May, pp. 270-278.

_____ and Lawrence M. DeBrock (1980)
"The Structural Effects of State Regulation of Retail Fluid Milk
Prices," *Review of Econ. and Stat.*, 62:254-262, May.

Mathis, Anthony G. (1958)
The Probable Impact of Milk Concentrates on the Fluid Milk Market,
U.S. Agric. Mktg. Serv., Mktg. Res. Rpt. No. 208, Feb., 24 pp.

_____, G.B. McIntyre, and J. Toomey (1965)
Government's Role in Pricing Milk in the United States, ERS-239,
May, 39 pp.

_____ and Dolores Fravel (1968)
Government's Role in Pricing Fluid Milk in the United States, U.S.
Econ. Res. Serv., Agric. Econ. Rpt. No. 152, Dec., 31 pp.

_____, D.E. Friedly, and S.G. Levine (1972)
Government's Role in Pricing Fluid Milk in the United States, U.S.
Econ. Res. Serv., Agric. Econ. Rpt. No. 229, July, 33 pp.

Matulich, Scott C. (1978)
"Efficiencies in Large-Scale Dairying: Incentives for Future Structural Change," *Am. Jnl. Agric. Econ.*, 60:642-647, Nov.

Mayer, Leo V. (1979)
"The Purpose, Function, and Policy Behind the Federal Government's
Role in Milk Marketing: 1929-1979," *Proceedings: Seminar on Examin-
ing Dairy Policy Alternatives*, Sept.

McGuire, David P., and Bernard F. Stanton (1979)
Are There Limits to Herd Size on New York Dairy Farms? Cornell A.E.
Res. 79-16, Aug., 27 pp.

Mehren, George L. (1947)
Agricultural Market Control Under Federal Statutes, Univ. of Calif.
Gianninni Found. of Agric. Econ., Mimeo Rpt. No. 90, Oct., 48 pp.

Merritt, Eugene (1915)
The Production and Consumption of Dairy Products, U.S. Dept. Agric.,
Bull. No. 177, Feb., 19 pp.

Metzger, Hutzel (1930)
Cooperative Marketing of Fluid Milk, U.S. Bu. of Agric. Econ., Tech.
Bull. No. 179, May, 91 pp.

Miller, James J. (1982a)
Fluid Milk Markets: Conceptual Models and Considerations Related to Stability, U.S. Econ. Res. Serv., forthcoming, 16 pp.

_____ (1982b)
Impact of Ultra-High Temperature Milk on the U.S. Dairy Industry, Ph.D. thesis, Purdue Univ., Dec., 108 pp.

Miller, Robert R. (1969)
"Regional Changes in Dairy Manufacturing," *Dairy Situation,* U.S. Econ. Res. Serv., DS-328, Nov. pp. 34-44.

_____ (1971)
"Developments and Trends in the Casein Market," *Dairy Situation,* U.S. Econ. Res. Serv., DS-334, March, pp. 29-38.

Moede, Herbert H. (1971)
Over-the-Road Costs of Hauling Bulk Milk, U.S. Econ. Res. Serv., Mktg. Res. Rpt. No. 919, Jan., 26 pp.

Mueller, Willard F., Larry G. Hamm, and Hugh L. Cook (1976)
Public Policy Toward Mergers in the Dairy Processing Industry, N.C.-117 Monograph 3, North Central Reg. Res. Pub. 233, 120 pp., Dec.

Mumford, Herbert W., Jr. (1934)
"Day-of-the-Week Variation in Sales of Milk and Cream to Stores," *Farm Economics,* Cornell Univ., No. 84, Feb., pp. 2036-38.

Murphy, Paul L. (1955)
"The New Deal Agricultural Program and the Constitution," *Agric. History,* 29:160-169, Oct.

National Comm. for the Review of Antitrust Laws and Procedures (1979)
Report to the President and the Attorney General of the National Commission, Jan. 16.

National Commission on Food Marketing (1966)
Organization and Competition in the Dairy Industry, Tech. Study No. 3, June, 409 pp.

New York Crop Reporting Service,
New York Dairy Statistics, annual.

North Central Regional Dairy Marketing Research Committee (1953)
Outer-Market Distribution of Milk in Paper Containers in the North Central Region, Purdue Univ. Agric. Expt. Sta. Bull. 600, Oct., 44 pp.

Nourse, Edwin G. (1935)
Marketing Agreements Under the AAA, The Brookings Inst., 446 pp.

_____, et al (1962)
Report to the Secretary of Agriculture by the Federal Milk Order Study Committee, produced through the facilities of the U.S. Dept. of Agric., Dec., 105 pp.

Nicholls, William H. (1939)
Post-War Developments in the Marketing of Butter, Iowa Agric. Expt. Sta., Res. Bul. 250, Feb., 44 pp.

Novakovic, Andrew, and Robert L. Thompson (1977)
"The Impacts of Imports of Manufactured Milk Products on the U.S. Dairy Industry," *Am. Jnl. Agric. Econ.,* 59:507-19, Aug.

_____ and Richard Aplin (1982)
An Analysis of the Impact of Deregulating the Pricing of Reconstituted Milk under Federal Milk Marketing Orders, Cornell Agric. Econ. Staff Paper No. 82-6, March, 12 pp.

O'Connell, P.E. (1954)
Nonfarm Consumption of Fluid Milk and Cream. A Revised Series of Statistics (United States, 1924-50: Urban Markets, 1944), U.S. Agric. Mktg. Serv., Mktg. Res. Rpt. No. 72, May, 51 pp.

O'Day, Lyden (1978)
Growth of Cooperatives in Seven Industries, U.S. Econ., Stat., and Coop. Serv., Coop. Res. Rpt. 1, July, 50 pp.

Parker, Russell W. (1973)
Economic Report on the Dairy Industry, Staff Report to the Federal Trade Commission, March 19 (Released Nov. 20, 1974), 152 pp.

Pearson, F.A. (1919)
"Principles Involved in Fixing the Price of Milk," *Jnl. Farm Econ.,* 1:89-96, Oct.

Persigehl, Elmer S., and James D. York (1979)
"Substantial Productivity Gains in the Fluid Milk Industry," *Monthly Labor Review,* Bu. of Labor Stat., July, pp. 22-27.

Peterkin, Betty, Judy Chassy, and Richard Kerr (1975)
The Thrifty Food Plan, U.S. Agric. Res. Serv., Sept., 26 pp.

Pollard, A.J. (1938)
Transportation of Milk and Cream to Boston, Vt. Agric. Expt. Sta., Bull. 437, June, 42 pp.

Rada, Edward L., and D.B. Deloach (1941)
An Analysis of State Laws Designed to Effect Economic Control of the Market Milk Industry, Oreg. State Coll., Studies in Econ. 2, Dec., 72 pp.

Roadhouse, Chester L., and James L. Henderson (1950)
The Market-Milk Industry, McGraw-Hill Book Co., Inc., 716 pp.

Roberts, Tanya (1975)
A PIE-C Reseach Report: Review of Economic Literature on Milk Regulation, Public Int. Econ. Ctr., Dec., 66 pp.

Rojko, Anthony S. (1957)
The Demand and Price Structure for Dairy Products, U.S. Agric. Mktg. Serv., Tech. Bull. No. 1168, 252 p.

Rourke, John P. (1982)
"Fluid Milk Processor Structure in Federal Order Markets," *Federal Milk Order Statistics for March 1982,* U.S. Agric. Mktg. Serv., FMOS-267, May, pp. 39-46.

Russell, Howard S. (1976)
A Long, Deep Furrow. Three Centuries of Farming in New England, Univ. Press of New England, 672 pp.

Salathe, Larry, William D. Dobson, and Gustof Peterson (1977)
"Analysis of the Impact of Alternative U.S. Dairy Import Policies," *Am. Jnl. Agric. Econ.,* 59:496-506, Aug.

Scanlan, John J. (1937)
An Economic Study of the Transportation of Milk in the Philadelphia Milkshed in Relation to the Operations of the Interstate Milk Producers' Association, Inc., U.S. Farm Credit Admin., Bull. No. 13, April, 151 pp.

Schlesinger, Arthur M., Jr. (1959)
The Age of Roosevelt. The Coming of the New Deal, Houghton Mifflin Co., 669 pp.

Schnittker, John, Emerson M. Babb, Jerome W. Hammond, Ronald Knutson, H. Alan Luke, and Robert Story (1979)
Evaluating Dairy Policy Alternatives, Milk Ind. Found., 104 pp., Sept.

Scott, Forrest E., et al (1957)
Farm-Retail Spreads for Food Products, Costs, Prices, U.S. Agric. Mktg Serv., Misc. Pub. No. 741, Nov., 165 pp.

Shaffer, James D. (1979)
"Observations on the Political Economics of Regulations," *Am. Jnl. Agric. Econ.,* 61:721-731, Nov.

Shaw, Charles N., M.C. Hallberg and C.W. Pierce (1975)
Impacts of State Regulation on Market Performance in the Fluid Milk Industry, Penn. Agric. Expt. Sta., Bull. 803, July, 39 pp.

_____ and S.G. Levine (1978)
Government's Role in Pricing Fluid Milk in the United States, U.S. Econ., Stat., and Coop. Serv., Agric. Econ. Rpt. No. 397, March, 40 pp. (also revised table 5 dated May 4, 1978).

Smith, Art, and Ronald D. Knutson (1979)
Economic Formula Pricing of Milk, Texas Agric. Mktg. Res. and Dev.
Ctr., Res. Rpt. MRC 79-1, June, 32 pp.

Sommer, Hugo H. (1946)
Market Milk and Related Products, Madison, Wis., 745 pp.

Song, Dae Hee, and M.C. Hallberg (1980)
Are Dairy Programs Biased Toward Producers or Consumers? Penn State
Univ., Dept. of Agric. Econ. & Rur. Soc., Staff Paper No. 38,
Nov., 28 pp.

_____ and _____ (1982)
"Measuring Producers' Advantage from Classified Pricing of Milk,"
Am. Jnl. Agric. Econ., 64:1-7, Feb.

Sonley, L.T. (1940)
Cost of Transporting Milk and Cream to Boston, Vt. Agric. Expt.
Sta., Bull. 462, July, 56 pp.

Spencer, Leland (1929)
*An Economic Study of the Collection of Milk at Country Plants in
New York,* Cornell Agric. Expt. Sta., Bull. 496, June, 47 pp.

_____ (1932)
"Milk Distribution in the Larger Cities of the United States," *Farm
Economics,* Cornell Univ., No. 74, Feb., pp. 1706-1708.

_____ (1933)
"Practice and Theory of Market Exclusion Within the United States,"
Jnl. Farm Econ., XV:141-158, Jan.

_____ (1936)
American Creamery and Prod. Review, June.

_____ (1938)
The Surplus Problem in the Northeastern Milksheds, U.S. Farm Credit
Admin., Bull. 24, April.

_____ (1941)
The Dairy Farmers Strike, Cornell Univ., A.E. 353, July, 10 pp.

_____, H.R. Kling, and I.R. Bierly (1941)
The Distribution of Milk by Subdealers in New York City, Cornell
Univ., Agric. Expt. Sta., A.E. 358, Sept., 13 pp.

_____ (1951)
Some Effects of Federal and State Regulation of Milk Prices, Cornell
Univ., Bull. A.E. 727, Jan., 42 pp.

Printed and bound by CPI Group (UK) Ltd, Croydon, CR0 4YY

23/10/2024

01778244-0007

_____ and S. Kent Christensen (1954)
Milk Control Programs of the Northeastern States. Part 1. Fixing of Prices Paid and Charged by Dealers, Cornell Univ. Agric. Expt. Sta., Bull. 908, Northeast Regional Pub. No. 21, Nov., 136 pp.

_____ and _____ (1955)
Milk Control Programs of the Northeastern States. Part 2. Administrative and Legal Aspects, and Coordination of State and Federal Regulation, Cornell Univ. Agric. Expt. Sta., Bull. 918, Northeast Reg. Pub. No. 23, Nov., 128 pp.

_____ (1972)
Federal Regulation of Milk Prices in World War I, Cornell Univ. Agric. Expt. Sta., Bull. A.E. Res. 72-13, 90 pp.

_____ and Charles J. Blanford (1973)
An Economic History of Milk Marketing and Pricing: A Classified Bibliography with Reviews of Listed Publications, 1840-1970, Columbus, Ohio: Grid Publ. Co., 978 pp.

Sprague, Gordon W. (1940)
Butter Price Quotations at Chicago, Ph.D. thesis, Univ. of Minnesota, March, 159 pp.

Stiebeling, Hazel K., Sadye F. Adelson, and Ennis Blake (1942)
Low-Priced Milk and the Consumption of Dairy Products Among Low-Income Families, 1940, U.S. Dept. of Agric., Circ. 645, 28 pp.

Stitts, T.G., J. J. Scanlan, H. J. Limbacher, and J. H. Lister (1933)
Report on the Survey of Milk Marketing in the Northeastern States, U.S. Farm Credit Admin. in cooperation with National Cooperative Milk Producers Federation and U.S. Dept. of Agric., July.

Stocker, Noel (1954)
Progress in Farm-To-Plant Bulk Milk Handling, U.S. Farmer Coop. Serv., Circ. 8, Nov., 53 pp.

Strauss, Frederick, and Louis H. Bean (1940)
Gross Farm Income and Indicies of Farm Production and Prices in the United States, 1869-1937, U.S Bu. of Agric. Econ., Tech. Bull. No. 703, Dec.

Sullivan, William G. (1942)
The Relief Milk-Distribution Program, U.S. Dept. Agric., Nov., 65 pp.

Taylor, George R., Edgar L. Burtis, and Frederick V. Waugh (1939)
Barriers to Internal Trade in Farm Products, U.S. Bu. Agric. Econ., March, 104 pp.

Tilsworth, Robin, and Truman Graf (1981)
"The Economic Impact of Imitation Cheese," *Economic Issues,* No. 60, Univ. of Wis.-Madison, July, 4 pp.

Tinley, James M. (1938)
Public Regulation of Milk Marketing in California, Univ. of Calif.
Press, 213 pp.

Trelogan, Harry C., and French M. Hyre (1939)
Cooperative Marketing of Dairy Products, U.S. Farm Credit Admin.,
Circ. No. C-116, June, 47 pp.

Trout, G.M., Carl W. Hall, and A.L. Rippen (1961)
*Survey of Homogenized Milk in Michigan. I. The Extent, Operating
Conditions, and Utilization of Returns,* Article 43-67, reprinted
from the *Quarterly Bulletin* of the Mich. Agric. Expt. Sta.,
43:618-633, Feb.

Tucker, George C. (1972)
Need for Restructuring Dairy Cooperatives, U.S. Farmer Coop. Serv.,
Serv. Rpt. No. 125, July, 54 pp.

_____, William J. Monroe, and James B. Roof (1977)
Marketing Operations of Dairy Cooperatives, U.S. Farmer Coop. Serv.,
FCS Research Rpt. 38, June, 46 pp.

_____, James B. Roof, and William J. Monroe (1979)
Future Structure and Management of Dairy Cooperatives, U.S. Econ.,
Stat., and Coop. Serv., Farmer Coop. Res. Rpt. 7, July, 60 pp.

U.S. Agric. Adj. Admin. (1935)
*Agricultural Adjustment in 1934. A Report of Administration of the
Agricultural Adjustment Act, February 15, 1934, to December 31,
1934,* 456 pp.

_____ (1936)
*Agricultural Adjustment, 1933-35. A Report of Administration of
the Agricultural Adjustment Act, May 12, 1933, to Dec. 31, 1935,*
322 pp.

_____ (1937a)
*Agricultural Conservation, 1936. A Report of the Activities of the
Agricultural Adjustment Administration,* 200 pp.

_____ (1937b)
A Survey of Milk Marketing in Milwaukee, Mktg. Inf. Ser. DM-1, 119
pp.

_____ (1939a)
*Agricultural Adjustment, 1937-38. A Report of the Activities Car-
ried on by the Agricultural Adjustment Administration,...from Jan.
1, 1937, through June 30, 1938,* 385 pp.

_____ (1939b)
Report of the Associate Administrator of the Agricultural Adjustment Administration, in charge of the Division of Marketing and Marketing Agreements, and the President of the Federal Surplus Commodities Corporation, 1939, Oct., 67 pp.

_____ (1939c)
Agricultural Adjustment, 1938-39. A Report of the Activities Carried on by the Agricultural Adjustment Administration, July 1, 1938, through June 30, 1939, 142 pp.

U.S. Agric. Mktg. Admin. (1942)
Report of the Administrator of the Agricultural Marketing Administration, 1942, 83 pp.

U.S. Agric. Mktg. Service (1955)
Regulations Affecting the Movement and Merchandising of Milk. A Study of the Impact of Sanitary Requirements, Federal Orders, State Milk Control Laws, and Truck Laws on Price, Supply, and Consumption, Mktg. Res. Rpt. No. 98, June, 124 pp.

_____ (1959)
"Role of Governments in Pricing Fluid Milk in the United States," *Dairy Situation,* DS-273, August, pp. 21-48.

U.S. Bu. of Agric. Econ. (1942)
The Mktg. and Trans. Situation, July.

_____ (1943)
The Dairy Situation, DS-140, April.

_____ (1950)
Changes in the Dairy Industry, United States, 1920-50, A Statement Submitted July 21, 1950...before a Subcommittee of the Senate Committee on Agric. and Forestry. Reprinted from Utilization of Farm Crops. Milk and Dairy Products, Part 4, Appendix A, pp. 1963-2076.

_____ (1953)
"Fewer Plants Now Manufacturing Dairy Products, Larger Output Per Plant," *The Mktg. and Trans. Situation,* MTS-109, April, pp. 16-22.

U.S. Bu. of Labor Statistics (1951)
Fresh Milk Marketing in Large Cities, 84 pp.

_____ (1977)
Consumer Expenditure Survey: Diary Survey, July 1972-June 1974, Bull. 1959, 387 pp.

_____ (1979)
Productivity Indexes for Selected Industries, 1979 Edition, Bulletin 2054, 190 pp.

310

U.S. Bu. of the Census (1929)
Census of Business. Retail and Wholesale Trade.

_____ (1957)
Concentration in American Industry, Report prepared for the Subcom-
mittee on Antitrust and Monopoly, Committee on the Judiciary, U.S.
Senate, 85th Congress, 1st Session, 756 pp.

_____ (1966)
Concentration Ratios in Manufacturing Industry, 1963, 252 pp.

_____ (1975a)
Concentration Ratios in Manufacturing, 1972 Census of Manufactures
Special Rpt. Series, MC72(SR)-2, 244 pp.

_____ (1975b)
Historical Statistics of the United States, Colonial Times to 1970,
1231 pp.

_____ (1978)
Statistical Abstract of the United States, 1977, Sept., 1048 pp.

U.S. Dept. of Justice (1977)
Milk Marketing. A Report of the Task Group on Antitrust Immunities,
Jan., 77 pp.

U.S. Economic Research Service (1962a)
Government's Role in Pricing Fluid Milk in the United States,
ERS-63, Sept., 30 pp.

_____ (1962b)
Dairy Situation, Dec.

_____ (1975)
The Impact of Dairy Imports on the Dairy Industry, Agr. Econ. Rpt.
278, Jan., 77 pp.

U.S. Economics and Statistics Service (1981)
*U.S. Casein and Lactalbumin Imports: An Economic and Policy
Perspective,* ESS Staff Rpt. AGESS810521, June, 77 pp.

U.S. Economics, Statistics, and Cooperatives Service (1979)
"Plants Manufacturing Dairy Products, by Production-Size Groups and
Major Regions, United States, 1977" in *Dairy Products,* Da 2-6
(11-79), Nov., pp. 13-17.

U.S. Federal Surplus Commodities Corporation
Annual Reports, 1934-45.

U.S. Federal Trade Commission (1921)
*Report of the Federal Trade Commission on Milk and Milk Products,
1914-1918,* June 6.

_____ (1936)
Report of the Federal Trade Commission on the Sale and Distribution of Milk and Milk Products Through Certain Farm Cooperatives in the New York Metropolitan Area and by Nation-Wide Distributors, 7th Congress, 1st Session, House Document No. 95, 172 pp.

_____ (1937)
Report...on Agricultural Income Inquiry. Part I. Principal Farm Products, House Document 152, 74th Cong., 1st Session, 1134 pp.

U.S. National Resources Committee (1939)
The Structure of the American Economy. Part 1. Basic Characteristics, June, 396 pp.

U.S. Public Health Service (1963)
Utilization of the Milk Ordinance and Code Recommended by the U.S. Public Health Service, U.S. Dept. of Health, Education and Welfare, P.H.S. Pub. No. 1018, January, 29 pp.

U.S. Senate (1961)
Price Supports for Perishable Products: A Review of Experience. A Staff Report on the Scope and Cost of Price-Support Programs for Perishable Agricultural Commodities, 1933 to Date, October 17, 82d Cong., 1st Session, 45 pp.

U.S. Statistical Reporting Service (1974)
Number of Plants Manufacturing Dairy Products by Production-Size Groups, 1972, in *Dairy Products,* Da (2-74), Feb. 28, pp. 16-20.

_____ (1977)
Minnesota Dairy Statistics, 19 pp.

U.S. Surplus Marketing Administration (1941)
Report of the Administrator of the Surplus Marketing Administration, 46 pp.

U.S. War Food Administration (1943)
Report of the Director of the Food Distribution, Oct., 118 pp.

_____ (1944a)
Food Program for 1944, Feb., 96 pp.

_____ (1944b)
Report of the Director of the Office of Distribution, 1944, Oct., 118 pp.

_____ (1945)
Final Report of the War Food Administrator, 1945, June, 39 pp.

Vertrees, James G., and Peter M. Emerson (1979)
Consequences of Dairy Price Support Policy, Congressional Budget Office, March, 80 pp.

312

Vial, Edmund E. (1940)
Production and Consumption of Manufactured Dairy Products, U.S Bu.
of Agric. Econ., Tech. Bull. No. 722, April, 76 pp.

_____ (1977)
*Milk Production, Utilization and Prices with Milk Orders and Price
Supports,* in *The Market Administrator's Bulletin,* New York-New
Jersey Milk Mktg. Area, March, 73 pp.

Vosloh, Carl (1971)
The Formula Feed Industry in 1969, A Preliminary Report, U.S. Econ.
Res. Serv., ERS-494, Nov.

_____ (1978)
*Structure of the Feed Manufacturing Industry, 1975. A Statistical
Summary,* U.S. Econ., Stat., and Coop. Serv., Stat. Bull. No. 596,
155 pp.

Waite, Warren C., Don S. Anderson, and R.K. Froker (1941)
*Public Pricing of Milk, Part II of Economic Standards of Government
Price Control,* Donald A. Wallace, editor, U.S. Temporary National
Economic Committee, Monograph No. 32, 514 pp.

Waugh, Frederic V. (1964)
Demand and Price Analysis. Some Examples from Agriculture, U.S.
Econ. Res. Serv., Tech. Bull. No. 1316, Nov., 94 pp.

_____ (1967)
Factor Analysis: Some Basic Principles and an Application, in
*Level-of-Living Indexes for Counties of the United States, 1950,
1959, and 1964,* by John M. Zimmer and Elsie S. Manny, U.S. Econ.
Res. Serv., Stat. Bull. No. 406, June, pp. 67-70.

Whey Products Institute
A Survey of Utilization and Production Trends, annual.

Whitaker, G.M. (1898)
The Milk Supply of Boston and Other New England Cities, U.S. Dept.
Agr., Bur. Anim. Ind., Bull. 20, 37 pp.

_____ (1905)
The Milk Supply of Boston, New York, and Philadelphia, U.S. Dept.
Agric., Bur. Anim. Ind., Bull. 81, 62 pp.

Williams, Sheldon W. (1941)
*Studies in Vermont Dairy Farming. XII. Dairy Farm Management in
the Champlain Valley and Its Relation to the Price Level,*
Vermont Agric. Expt. Sta., Bul. 479, June, 276 pp.

_____ (1951)
Marketing Dairy Products, paper at the Southern Agricultural Workers
Conference, Memphis, Tenn., Feb.

_____, R.W. Bartlett, E.F. Baumer, T.F. Graf, E.F. Koller,
G. McBride, and J.B. Roberts (1962)
*The Mechanics of Supply-Demand Adjusters for Midwestern Milk Mar-
kets,* N.C. Reg. Pub. 134, IL Ag. Expt. Sta. Bull. 684, April, 62 pp.

_____, David A. Vose, Charles E. French, Hugh L. Cook,
and Alden C. Manchester (1970)
Organization and Competition in the Midwest Dairy Industries, Iowa
State Univ. Press, 339 pp.

Index